SMALL COMPUTERS IN CONSTRUCTION

Proceedings of a symposium sponsored by
the ASCE Construction Division in conjunction
with the ASCE National Convention
Atlanta, Georgia
May 14-18, 1984

Wayne C. Moore, Editor

Published by the
American Society of Civil Engineers
345 East 47th Street
New York, New York 10017-2398

The Society is not responsible for any statements
made or opinions expressed in its publications.

PREFACE

Wayne C. Moore, M. ASCE

The emergence of small computers in the business place is nothing short of revolutionary. For the construction industry this explosion of distributed, inexpensive computing capability holds the promise of increased efficiency and many other far reaching benefits. At the same time, for a contractor, engineer or other industry businessman trying to use this capability on a practical, day-to-day level, the options can be overwhelming. Use of small computers in the construction industry so far, has closely paralleled other business applications, primarily for accounting, word processing, and similar data processing functions. The unique nature of the construction process as we know it today, offers an array of potential applications that are limited only by the imagination of the user. The purpose of the Symposium is to communicate an overview of current developments and trends in the utilization of this new resource in the construction industry.

In October 1982, the ASCE Construction Division formed a Task Committee on the Application of Small Computers in Construction. Objectives of the Task Committee are to examine current developments and trends for the utilization of small computers in construction and to develop recommendations as to what actions and objectives, if any, ASCE's Construction Division should have in this area. The committee's membership was formed by open call and is a cross section of ASCE members representing construction contractors, owner/agency representatives, academics and consultants.

Membership of the Task Committee was polled to solicit areas of interest for papers to be presented at the 1984 ASCE Spring Convention in Atlanta. Suggestions submitted fell into three general categories:

1. Needs, Applications and Trends
2. Acquiring the ''Right'' system
3. Software

These categories became the session topics for the Symposium. A general Call for Abstracts was published in May 1983 and selection of papers for presentation was made by members of the Task Committee. The papers selected cover a spectrum of topics from futuristic concepts of robotics to estimating practicality. Each of the papers included in the Proceedings has been accepted for publication by the Proceedings Editor. All papers are eligible for discussion in *The Journal of Construction Engineering and Management*, and are also eligible for ASCE awards.

Sponsor: Task Committee on the Application of Small Computers in Construction of the Construction Division, ASCE.

Symposium Committee: Wayne C. Moore, Chairman, Boyd C. Paulson, Jr., Daniel W. Halpin

CONTENTS

Needs, Applications and Trends
Session No. 16-S2.1

Acquiring the "Right" System
Session No. 23-S2.2

Software
Session No. 29-S2.3

*Manuscript not available at time of printing.

*Manuscript not available at time of printing.

Trends in Small Computer Utilization

By Lansford C. Bell[1], M. ASCE, Garabet Bostajian[2] and Mary Manson[3]

Introduction

An electronic digital computer is a device that is capable of performing detailed mathematical and logical calculations, storing and manipulating vast amounts of data, and printing or displaying information in a meaningful format. Recent innovations in computer technology have resulted in the development of inexpensive yet extremely powerful microprocessor-based digital computer systems. These computers are generally called professional desktop microcomputers if they are small, single user systems that (unlike so-called personal computers) are adequate for a business environment and cost between $5,000 and $15,000. The term minicomputer is then applied to systems with a considerably greater processing and storage capability, a multiuser potential, and a minimum system configuration cost of $15,000 to $75,000.

A tremendous volume of technical literature now exists that sets forth specific guidelines for evaluating and acquiring small microprocessor-based computers (microcomputers, minicomputers) for engineering and general business applications (2, 8, 9). Reports, articles, short courses and newsletters that address the unique computer requirements of construction contractors have also become fairly common in the last few years (3, 5, 6). This literature generally deals with the very important topics of software evaluation, hardware capabilities and options, contract RFPs and the role of consultants, etc. Few if any of these publications have been based on a substantial amount of feedback from construction contractors that have actually purchased and implemented microprocessor-based computers as their primary means of data processing.

In an effort to determine the extent to which micro and minicomputers have been successfully implemented by construction contractors, the authors compiled a comprehensive questionnaire that was mailed in April 1982 to 435 random contractors throughout the Southeastern United States. One hundred twenty four contractors completed the questionnaire survey form that requested information pertaining to the size of the contractors firm (annual dollar volume, number of W2 forms processed, etc.), the type of computer system or service bureau being

---

[1] Associate Professor, Department of Civil Engineering, Auburn University, Auburn, Alabama 36849.

[2] Consulting Civil Engineer, Sharjah, U.A.E.

[3] Graduate Research Assistant, Department of Civil Engineering, Auburn University, Auburn, Alabama 36849.

1

utilized and the type of data processing applications that have been implemented. The questionnaire also asked the contractor to rate the prewritten application software packages in current use by his firm and to identify the benefits associated with a small computer purchase.

The April 1982 questionnaire survey was updated in December 1983 by contacting 67 of the original survey participants by telephone. A cross section of computer users, non-users and service bureau subscribers were asked to comment on recent changes in their hardware or software utilization.

Data obtained from the questionnaire survey and the telephone update form the basis of the overall trends described herein. Additional information was obtained by attending various hardware vendor demonstrations, computer use seminars and software vendor training sessions.

Questionnaire Survey

Microcomputer vs. Minicomputer. - As stated above, 124 responses were received from a comprehensive questionnaire survey that was conducted in April 1982. The general distribution of those questionnaire responses is shown in Table 1. Thirty-two of the fifty-seven contractors (56%) that reported owning their own computer had purchased their computer since January 1, 1980.

Table 1. - Data Processing Modes Used by Surveyed Contractors

Principal Data Processing Mode* (1)	Number of Surveyed Contractors (2)	Number of Surveyed Contractors (3)
In-House Microcomputer	18	14%
In-House Minicomputer	32	26%
In-House Mainframe Computer	7	6%
Service Bureau	15	12%
No Computer	52	42%
	124	100%

*Five contractors surveyed used both in-house computers and a service bureau. Their principal data processing mode was assumed to be the in-house computer.

It should be noted that the contractors participating in the survey were selected at random from contractor association mailing lists and telephone directory listings. Responses were therefore received from a high percentage of small specialty contractors and home builders. The general distribution of computer classifications as a

function of annual contract volume and number of W2 forms processed is
shown in Figure 1. Contractors reporting no in-house computer averaged
$2.3 million in annual contracts, whereas contractors reporting a micro-
computer or a minicomputer as their primary vehicle for electronic data
processing averaged $2.0 million and $12.3 million in annual contracts
respectively.

The number of annual W2 forms processed by a construction firm
may be a more reliable indicator of actual data processing requirements
than annual contract volume. The average number of W2 forms processed
by microcomputer users was 55; the average number of W2 forms processed
by minicomputer users was 450.

It is interesting to note that contractors with annual work volumes in excess of $5 million or annual W2 volumes in excess of 100
have chosen mini as opposed to microcomputers. Minicomputer users
reported annual contract volumes ranging from $2 million to $58 million,
with the number of permanent employees ranging from 14 to 300 and the
number of W2 forms processed annually ranging from 75 to 2500.

Hardware Preferences. - A wide range of hardware brands, capabilities and options were reported in the April 1982 survey by the contractor users. The most common microcomputer reported at that time was
the Radio Shack TRS-80 Model II. Most users were using this computer
with a 64K RAM, 2 or 3 optional floppy disk drive units, and a 160 CPS
dot matrix printer. Other popular micros included the Apple II and the
IBM Datamaster.

In the December 1983 telephone survey update, a number of contractors reported replacing their floppy disk drives with hard disks
and a substantial number had purchased IBM microcomputers.

The most popular minicomputer reported by the survey was the
Digital Equipment Corporation (DEC) PDP 11 Series with 128K to 256K
RAM, 10M or more hard disk mass storage capability, and a 180 CPS
printer. The IBM System 34 minicomputer and the older and no longer
marketed IBM System 32 were also popular hardware configurations.
Radio Shack, DEC and IBM hardware are apparently popular in the Southeastern United States because of the availability of prewritten applications software that is compatible with these vendor's products.

Time and Cost Savings. - Contractors participating in the questionnaire survey were asked to comment on the direct savings in either
time or money that resulted from their computer purchase. Many contractors commented on time savings that resulted from computer ledger
postings and computer payroll preparation. Savings of three or four
days per month were cited. Most of the contractors surveyed agreed,
however, that dollar savings could not be directly determined. They
also agreed that because reports could now be processed in an accurate
and timely manner, particularly in the area of job cost analysis, the
computer purchase was more than justified.

Software Utilization. - As part of the questionnaire survey described above, the participating construction contractors were asked to

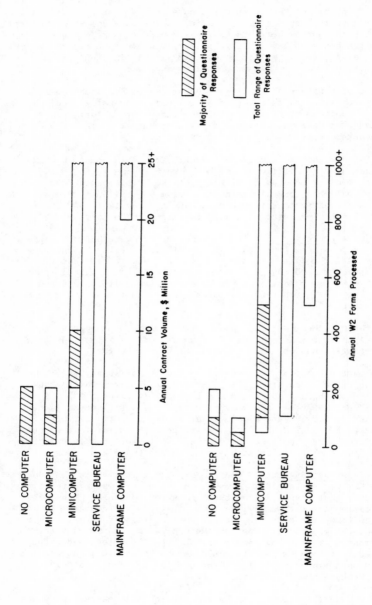

Figure 1. Distribution of 124 Responses to a Computer Utilization Questionnaire.

identify and describe the application software they had purchased, leased or developed in-house. Thirty-nine of the fifty-seven contrac-- tors (68%) that reported owning their own computer had purchased or leased comprehensive accounting software packages that had been developed by software vendors specifically for the construction indus- try. This type of accounting software generally consists of four or more integrated modules or programs that perform the basic accounting functions.

The term "integrated" is applied to software packages that have modules which automatically transfer data and program output between modules. Integrated packages are therefore highly desirable because they eliminate the necessity of inserting the same data for different programs or manipulating the output of one program to serve as input for a different program. In an integrated system, accounts payable and payroll transactions are posted automatically to job cost and general ledger files. Estimating modules in integrated systems may or may not interact with the modules used for tracking job cost.

Practically all microcomputer users participating in the survey purchased their software from independent third party vendors. The cost of such software ranges from $1000 to $3000 per module with some vendors charging extra for training and support. Microcomputer owners that had purchased general purpose business software from their hard- ware vendor found such programs totally inadequate for construction applications.

In most cases, minicomputer users reported purchasing their soft- ware and hardware at the same time in one "turnkey" package. Under the turnkey arrangement, a single vendor sells, services, updates and supports the entire hardware/software system. The cost for a minimum configuration turnkey system consisting of a single user 128K mini- computer and four integrated software modules was reported to be between $35,000 and $50,000. Minicomputer systems were occasionally purchased after asking vendors to respond to a formal proposal request. Consultants were sometimes employed to prepare the proposal requests.

Construction contractors participating in the questionnaire survey were also asked to subjectively rate their application software prog- rams. A summary of their ratings is shown in Table 2. It is somewhat surprising to see 90% of the contractors utilizing their computers for accounting purposes but only 31% using the computer for estimating and 10% using the computer for project scheduling.

For the most part, both mini and micro users were well satisfied with the software packages that were written either by outside soft- ware vendors or by their own company personnel. Some contractors were critical of the constrained report formats and the limiting capacity parameters of some of the prewritten modules, however. A more detailed analysis of the questionnaire responses appears in Reference 1.

Survey Update

In December 1983, 68 of the original 124 survey respondents were con- tacted by telephone. At the time of the original April 1982 survey,

TABLE 2. - Software Ratings of Surveyed Contractors With In-House Computers

Software Module (1)	Contractors With In-House Computers That Utilize The Module (Percent)* (2)	Ratings of Contractors Utilizing Commercially Prepared Software Modules (Percent)			
		E (3)	G (4)	A (5)	P (6)
General Ledger	91	55	38	7	0
Accounts Payable/ Receivable	91	34	54	10	2
Payroll	91	48	40	12	0
Job Cost	89	33	46	18	3
Estimating	31	43	14	29	14
Scheduling	10	0	60	40	0

*Includes software modules written by in-house personnel.

41 of this group of 68 contractors owned a small computer, 12 were utilizing service bureaus, 12 were non-computer users, and 3 were using service bureaus and an in-house computer.

Eight of the 41 computer owners had significantly altered their hardware configurations (primarily to expand mass storage) and four had purchased new portable micros (to assist with estimating and bid preparation). A number of firms reported increased utilization of word processing and spreadsheet software but few new application program modules, other than those associated with the traditional accounting function, had been purchased.

Ten of the 12 service bureau contractors were still using the same service bureau with no complaints. Three of the 12 contractors that were not using computers in April 1982 had made a computer purchase by December 1983 and another 5 were seriously considering such a purchase.

The 41 contractors that had been using small computers since April 1982, or before, were asked to comment on staff training and computer education within their firm. Most contractors reported no difficulty in training employees for computer use but some criticism of user manuals was expressed.

In the survey update the contractor computer users were also asked if they were using their computers for such unique applications as computer-aided drafting or design, simulation, risk analysis, robotics, networking information systems, etc. Other than the three firms that are using their micros to communicate with a home office mainframe computer, no unique or innovative applications were reported. Very few contractors had any experience with such recent hardware innovations as the mouse, light pen or touch sensitive screen. Similarly, few contractors were aware of the capabilities of the newer integrated software packages (Lotus 1-2-3, Lisa software by Apple and others) that are capable of sharing or moving data between individual program modules. Two contractors commented that the economic climate of the last two years had effectively discouraged computer expansion and experimentation.

Conclusion

Construction contractors with annual contract volumes of about $50 million or less are successfully utilizing micro and minicomputer as their primary means of data processing. Basic integrated accounting software modules that are being marketed by third party vendors and turnkey distributors appear to be adequate for most contractor's needs.

This is not to say, however, that all software on the market performs as it should. There are apparently far more variations in software quality and support than in hardware reliability and service. Furthermore, most quality software packages are compatible with only a single hardware system. Contractors are therefore very carefully evaluating the available software before purchasing their computer systems.

It is somewhat surprising to note the general reluctance of construction contractors to use their small computers for tasks outside the traditional accounting-estimating-job cost areas. As contractors become more computer oriented and construction software becomes more integrated and user friendly, innovative construction applications will perhaps become more commonplace.

Appendix I. - References

1. Bostajian, Garabet, "Selection and Utilization of Small Computers by Southeastern Construction Firms," Thesis presented to Auburn University, Auburn, Alabama, in 1982, in partial fulfillment of the requirements for the degree of Master of Science.

2. Brown, J.C., "Recommended Procedure for Computer Selection," Journal of the Technical Councils of ASCE, Vol. 105, No. TC2, Dec., 1979, pp. 319-326.

3. Carota, James D., "The Small Contractor and the Small Computer," Technical Report No. 248, Department of Civil Engineering, Stanford University, Stanford, California, June, 1980.

4. Construction Computer Application Directory, Construction Industry Press, Silver Spring, Md.

5. Construction Computer Applications Newsletter, Construction Industry Press, Silver Spring, Md.

6. Dunder, Vera D., "Computer Acquisition for the Small Contractor," Journal of the Construction Division, ASCE, Vol. 106, No. CO2, June, 1980, pp. 173-184.

7. Grobler, Francis, Michael J. O'Connor and Glenn E. Colwell, "Microcomputer Selection Guide for Construction Field Offices," Technical Report P-146, Construction Engineering Research Laboratory, U.S. Army Corps of Engineers, June, 1983.

8. Rooney, Martin F., "Computer Hardware for Civil Engineers," Journal of the Technical Councils, ASCE, Vol. 107, No. TC1, April 1981, pp. 153-168.

9. U.S. Department of Transportation, Urban Mass Transportation Administration, "Microcomputers in Transportation - Selecting A Single User System," Report UMTA-URT-41-83-4, April, 1983.

AUTOMATED CONTROL AND ROBOTICS FOR CONSTRUCTION

By Boyd C. Paulson, Jr., [1] M. ASCE

Abstract: This paper addresses the potential for automated process control and robotics for remote, large-scale field operations such as those on construction engineering projects. Classifications of technologies for automated process control and robotics in such operations include on-site process control for fixed plants; partial or fully automatic control of mobile equipment; fixed-base manipulators; mobile robots; communications between on-site computers and automated machinery; monitoring and data recording via on-board microprocessor-based sensors; electronic ranging and detection; and video-image pattern recognition. Combining selected technologies with microcomputer-based software could provide a means of coordinating various discrete automated components or machines that must work together to perform field tasks. This paper also mentions categories of needs for such technologies on field operations, and potential barriers to implementation. Progress will depend on the interest and support of researchers qualified to advance this field.

Introduction

Factory-based manufacturing industries have long been enjoying increasing economic, safety, and quality control benefits from automated process control of production operations, and robotics applications are being incorporated as a logical extension of this trend. However, field-oriented industries like construction, with frequently reconfigured operations and often severe environmental conditions, have been slower to adopt process control technologies, and scarcely have been touched by robotics, at least in the United States. This paper examines reasons for this situation, and suggests technologies and research that might do something about it.

---

[1] Professor and Associate Chairman, Department of Civil Engineering, Stanford University, Stanford, CA 94305.

Background and Overview

An April 28, 1983 Engineering News-Record (ENR) report on the 13th International Symposium on Robots stated, "Of the 260 robots on display, . . . not one was earmarked for construction applications." Furthermore, " . . . conference exhibitors were not aware of any current job-site applications in the U.S." [20, p. 30] Although construction is considered by some to be the largest sector of the U.S. economy, robot manufacturers gave several reasons for shunning this market. According to David M. Osborne, technical director of Swedish ASEA's Troy, Michigan office, "Construction jobs are not always the same, so there's not a great deal of repeatability. Most construction jobs require a certain amount of on-site judgment, which a robot can't provide. And there are a lot of uncontrolled environmental factors. If I leave a robot out in the rain, I lose $100,000 by the end of the week." [20, p. 30]

David Wisnoski, group vice president of Naperville Illinois-based /Industrial Systems Group, commented that, "Where robots work best is where the environment is very structured. The construction industry . . . is very individualistic and resists this kind of thing. I wouldn't say people should be worried about their jobs yet - there won't be robots in their business for quite some time." [20, p. 30]

As outsiders to construction, these robot manufacturers are quite perceptive in identifying the major technical and institutional issues that face field and project-oriented industries. It is surprising, however, that robot manufacturers seem to be so easily discouraged rather than challenged by the problems they describe. First, robots by nature are programmable and thus should be adaptable to changing environments, rather than be restricted to the repeatable, very structured tasks desired by the present state of the art. Second, most construction jobs do indeed require more judgment than those in factories, but is this not then an area where the benefits of artificial intelligence research [1] might best emerge? Finally, the environment is certainly a factor. But if ASEA's robots cannot handle even a week of gentle rain, what must they think of a winter on Alaska's North Slope, platform operations in the North Sea, working 1000 feet underwater off the California coast, 120° F and sandstorms in the Middle East, or 100 stories up framing steel on a cold winter day in Chicago? In many of these situations, robots could provide welcome relief from the safety and health problems construction workers face every day. Construction machinery is manufactured to cope with such environments, so surely construction robots could be also.

In any case, the manufacturers are correct that there are many problems to be solved, both basic and developmental, before robotics and process control technologies evolve to the state where they will play a significant role for industries like this. Many of these problems are ones that can make exciting long-term basic research topics suitable for researchers, if only they can be made more aware of the needs and characteristics of such industries. This paper will address some of the present and future automated process control and robotics technologies that might solve some of the problems, mention the technical, economic, and institutional constraints of the construction industry that must be overcome for the technologies to prosper, identify potential applications on which construction could begin to benefit from such advances, and briefly comment on the state of the art in selected areas.

Objectives

It is important to emphasize that this is an _exploratory_ paper and is not intended significantly to report research results or describe applications of the technologies being examined. The general purpose is, given both the technologies and the industries in which they might eventually be applied, to build a bridge between them. More specific objectives are briefly to define, examine, and classify the various technologies; suggest experiments with both mobile and fixed-base robotics devices to better understand their potential and limitations; mention technical, economic, and institutional barriers to making automated real-time process control and robotics technologies an attractive goal and practical reality for industries like construction; and stimulate potential researchers with the qualifications and interests to advance this field.

Understanding and Classifying Potential Applications. In order for automated real-time process control and robotics to begin having a greater impact in construction, it will be necessary to acquire a sound understanding of current and projected technological developments in several discipline areas. Development of a useful classification system would help researchers define appropriate topics and objectives, and guide industry and government in developing effective, coordinated long-term programs of research. Possible general parameters for classifications include: the engineering and scientific disciplines needed for further research; their potential areas for application on various capital- and labor-intensive large-scale remote field operations; their present and potential state of advancement (i.e., conceptual, basic research, applied research, development needing only technology transfer, commercial product, etc.); and their likely application areas in field operations (e.g., fully or partially automating present fixed and mobile production equipment, new design approaches suitable for process automation, development of suitable robots and androids).

The types of automated process control and robotics applications and technologies that could be considered include, but need not be limited to:

1) **On-site automated process control for fixed plants,** such as concrete batch plants, reinforcing bar cutting and bending, pipe fabrication, carpentry shops, pre-cast concrete element fabrication, and aggregate crushing and screening plants. This type of application would be the most similar to factory automation, and the main focus would be on adaptations for rapid and economical set-up and take-down (e.g., design considerations for portable plants), a high-turnover and often inadequately trained labor force, provisions for severe environmental conditions, and

adaptability to a frequently changing product mix. Considerable progress already has been made by industry in some of these areas.

2) **Partial or full automation of mobile construction equipment,** including trucks ranging from light utility vehicles to large off-highway haulers; excavating and grading equipment such as scrapers, loaders, shovels, compactors and graders; materials hoisting equipment such as cranes and forklifts; and a wide range of specialty equipment such as paving machines, tunnel boring machines, construction railways, cableways, and pipelayers. There is some progress in a few of these, but compared to category 1 above, this application barely has been touched for real-time programmed control, let alone anything approaching robotics.

3) **Fixed-base or dimensionally constrained manipulators** (sometimes called robot arm manipulators) [4,8,18]. These are among the basic ingredients of factory automation, and have obvious applications in areas such as category 1 above. But what could be done by relocating such manipulators into traditional field positions? A later section will mention a few promising attempts, but the scope and economic potential for these and other possible manipulator applications is far greater.

4) **Mobile robots and androids** [21,22,28], including those with wheel, track-type and walking transporters. Although such robots often contain one or more manipulating arms like their fixed-base counterparts, they are distinguished by a high degree of mobility, normally unconstrained by tracks, guidewires or other fixed references; they have a wider variety of sensors (sonic, light, touch, temperature, etc.) to cope with changing and less predictable environments; they are often battery powered to move without external power sources; and they typically have more on-board computer power to allow a greater measure of independent programmed analyses of and responses to their environment and tasks. Androids, which are robot devices that attempt to approximate the physical form and some functions of humans, are included so that the term "robotics" can apply to a wider range of mobile devices, without regard to their physical shape or functionality.

Technologies for Research and Development. To a large extent, construction can get a head start in automated process control and robotics by assimilating and projecting current and future scientific and engineering research in numerous areas, and by focusing the most promising technologies on the needs of project operations. However, concurrent laboratory experimental efforts are also important for several reasons. First, they will provide hands-on experiences with some of the key technologies. Second, they will provide means to implement and test technologies in configurations that might model - on a smaller scale and in a simplified environment - situations more typical of field operations, and that have not adequately been considered in factory applications. Technologies that might be explored via experimental research for their suitability in automated process control and robotics could include, but need not be limited to:

1) Microprocessor-based sensors for direct installation on production equipment for monitoring and control (e.g., strain gauges, load cells, displacement counters, etc.)

2) Remote sensor technologies such as electronic ranging and detection, sonic and ultrasonic detection and measurement, touch sensors, and light sensors. A major application of these would be to keep track of moving vehicles.

3) Microcomputer-based process control, including both analog and digital input and output for switching and manipulation [10,12,15, 16].

4) Open-loop feedback control systems, including microcomputer-based sensors mounted on a vehicles, where the mobile device or robot needs communications and sensing to maintain its position within the tolerances of its specified tasks.

5) Closed-loop servo mechanisms such as might be employed on fixed-base manipulators, where direct electronic or mechanical feedback provides continuous monitoring and precise control of position. For example, programable servo-type robot arm manipulators could provide experience with such mechanisms and further explore the possibilities for fixed-base manipulators in on-site production operations.

6) Monitoring and coordinating communications technologies, including analog radio and digital telecommunications [9,23,24], optical (e.g., laser) data transmissions, and electronic ranging and detection (e.g., radar), between microcomputers and mobile robots or vehicles.

7) Programming and experimentation with microcomputer-based coordination of multiple processes to achieve a common task (e.g., robot supplies materials to the manipulator and helps manipulate parts for fabrication or assembly).

8) Video image pattern recognition and image processing for use by robots in materials handling, orientation and navigation [7]. A promising device is the direct digital array detector that is like a light-sensitive memory chip.

Identifying Needs for and Barriers to Implementation. In contemplating the future of such technologies, it will be necessary to define, classify, and determine general priorities of needs for automated process control and robotics in several areas. Categories could include large vs. small projects, labor-intensive vs. capital-intensive operations, industry sectors (buildings, civil works, process plants, housing), phases and technologies within projects (site work, foundations, structural, piping, electrical, etc.), and types of firms (design-construction, general contractor, specialty contractor, etc.). Where will industry's progress be slowed for not implementing new automation and robotics technologies? Where will costs and schedules be unnecessarily high and quality low?

It will also be important to consider potential industry barriers. In a field like automated process control and robotics, there are certainly some very real social and economic as well as technical barriers that must be identified and overcome or accommodated if research efforts are to succeed eventually in development and implementation [3]. Researchers who proceed with such studies should be made aware of these barriers. In brief, the obstacles to technological advances are many in construction, and relate more to institutional problems like craft, company and process fragmentation, risk and liability, codes and standards, than they do to purely technological or economic concerns. Through anticipation and careful planning, the barriers can be overcome, as they have been to some degree in foreign countries and in selected parts of U.S. construction [19].

Identifying Potential Researchers. An important goal of an exploratory study of a promising new area is to stimulate qualified researchers to focus their interests on this field. Since many of the technologies and needs being considered will cross discipline boundaries (civil and construction engineering, mechanical engineering, electronics, communications, and computer science), an effort must be made to identify researchers or organizations with the interests and abilities to work in various areas. There are already many university and government researchers in the U.S. construction research communities. Through organizations such as the ASCE Construction Research Council, one could assess their interests in doing the type of research described here. What is much less well known, however, is how much interest researchers working at the cutting-edge of the various technological areas being described have in expanding their concepts and goals to the unconstrained, frequently reconfigured and often harsh environment of remote large-scale field operations. In establishing contacts with such researchers to learn about their technologies, one should simultaneously probe their interests in and capabilities for this type of research.

Present State of the Art

Given years of research in the defense and aerospace industries, manufacturing, and related fields of science and engineering, there has been a vast amount of work done on automated real-time process control and robotics, even from extraterrestrial sources let alone remote sites on earth. Construction researchers must start by exploring these fields and evaluating their technologies and research efforts for their potential relevance to construction needs and objectives. However, rather than attempt to review the vast general body of knowledge in data acquisition, automated process control, and robotics, this section will focus primarily on related work in construction.

The paragraphs below briefly review the following main areas that relate most closely to this paper: (1) On-site automated process control for fixed plants; (2) partial or full automation of mobile equipment; (3) fixed-base or dimensionally constrained manipulators; (4) mobile robots and androids; and (5) sensing and communications technologies.

On-Site Automated Process Control. The most progress to date in automation for construction has been made in control of temporary on-site plants for batching concrete, bending reinforcing steel, and making pre-cast concrete elements. For example, in batch plants for higher volume, high-quality concrete production, such as those for nuclear power stations, a computer controls the selection, transport, weighing, charging and mixing of cement, sand, aggregates, water and admixtures for a batch that meets specified design criteria for a specific structural component, and simultaneously handles administrative reporting for delivery, quality and cost control. To the extent that more construction processes and components can be redesigned for prefabrication in plant-type facilities, whether on-site or off-site, a larger fraction of construction processes can benefit from automated process control of this type.

Automation of Mobile Equipment. Most major construction equipment manufacturers are experimenting with and even producing some machines that include on-board microprocessors for monitoring performance, for maximizing engine power and fuel economy, optimizing gear shifts, keeping loads within safe tolerances, etc. These applications are beneficial, but fairly limited relative to the real potential for partial or fully automated control of machines as whole units in the overall production process.

More dramatic is the application of automated excavation grade control using laser surveying equipment [25], combined with electro-hydraulic feedback control systems mounted on bulldozers, motor graders, scrapers, etc. In the 1970s some government agencies and construction contractors helped pioneer these applications of partial automation [26, 27]. In applications such as highway grading, constructing large parking lots, and canals, these techniques have reduced costs in some cases by over 80% and improved quality (e.g., achieving subgrade thickness tolerances of 2% versus 10% to 20% otherwise achieved on normal work). They also permit the substitution of lower cost machines (e.g., bulldozers for motorgraders) and lower-skilled operators while giving quality improvements. Japan's Komatsu tractor company has also done related work in remote controlled amphibious and submersible bulldozers for work in coastal areas and hazardous environments [2,13]. Also in Japan, full-face tunneling machines are being implemented that incorporate a variety of sensing and control mechanisms to optimize tunnel excavation.

The next step, moving to more fully programmed automation, seems obvious but is much more difficult. At this stage the machines in effect would become robots with some limited programmed intelligence for self-guidance and control over complex routes and surfaces, making decisions on what to do about obstacles, and responding (if only to stop) to unanticipated changes. Researchers at Carnegie-Mellon University have proposed this type of automation, and have generated considerable interest [14]. Automation of independent machines needs much more research before it becomes safe and practical, but the potential is there.

Spatially-Constrained Manipulators. Much of the publicity about robotics in manufacturing industries currently centers around multi-jointed robot arm and hand mechanisms that are either attached to a fixed base, or to a platform such as a gantry that covers a clearly defined and limited area. With such a well-prescribed three-dimensional frame of reference, they can be programmed for operations requiring high precision.

Although the construction environment is often loosely constrained and frequently reconfigured, there is still considerable potential here for this type of automation. The processes can be redesigned to fit the tools, and the tools can evolve to handle more flexibly a wider variety of processes. Carnegie Mellon University researchers have experimented with such a robot arm for unit-masonry construction [14]. Japan's Shimizu Construction Company has mounted such an arm on a mobile platform and is using it to apply sprayed insulation in building construction [11]. Other possible applications are in tunnels where operations (e.g., liner erection) are highly repetitive and fit into well-defined geometric constraints.

Mobile Robots and Androids. Apart from Shimizu's platform-mounted robot arm mentioned above, there have been few attempts at developing and applying mobile, walking-type robots and androids even in manufacturing, let alone construction. However, owing to the nature of project sites, such devices might have even greater potential in construction than in plant-based manufacturing. A number of possibilities have been explored, including a walking spider-like robot to assist in the clean-up at the Three-Mile Island nuclear plant, others for assembly of space platforms and work on other planets, and a promising six-legged robot with a high power-to-weight ratio [21], but there is no real prototype as yet for a general-purpose robot that might be flexible enough to be a general utility tool on field projects. This is an attractive area for research.

Sensing and Communications Technologies. For years various kinds of instruments have been installed on or ahead of construction operations for technical or safety reasons. Particularly good examples are geotechnical and structural instruments for monitoring tunnel and foundation excavations. A new and promising application is monitoring grouting operations for a dam foundation [17]. Furthermore, the electronic revolution has brought rapid advances in surveying equipment for measuring distances, angles, and volumes. Mobile radio communications have also been a part of construction projects at least since the 1940s, not only among project engineers and supervisors, but for foremen directing equipment operators, operators calling for fuel or repair assistance, signalmen talking to crane operators, etc. In other words, there is already considerable experience with equipment and technologies that could support machine data acquisition and control functions. Little yet has been done, however, to take advantage of this existing infrastructure for purposes of automated process control.

In summary, remote field project-oriented industries like construction have only taken a few, loosely related steps toward automated process control and robotics for field operations. Initially there will be a lot to be learned from aerospace research and plant-based manufacturing industries, but in the long run both the challenges and potential rewards are even greater in large-scale field operations.

Conclusion

Efforts to define and stimulate interest in a broad program of research in this area could have a major impact on the technology and productivity of the construction industry. Construction has lagged the U.S. economy as a whole in recent decades, and in the process has handicapped other industries for which it provides capital facilities and raw materials [5,6]. Focusing the attention of high-technology researchers on this industry requires considerable development of mutual understanding, and that in large measure will require initiative from the construction industry.

Not surprisingly, there are also national economic reasons for getting such research underway. For example, in a feature cover story titled "Japan Takes Early Lead in Robotics," the July 21, 1983 issue of ENR described cases where robots already are performing economically useful tasks in the field for Japanese construction contractors [11]. The research laboratories of major Japanese engineering contractors [19] (laboratories which have few if any counterparts in the United States) are making this a high priority field in their research. Perhaps if no significant research effort evolves in the United States, American contractors will be able to solve their problems by importing robotics and process-control machinery from overseas. This could become one more area of technological advancement that Americans could give up without really trying, but is this wise?

Currently the United States has the lead in related computer technologies, artificial intelligence research and even some robotics technologies. What is needed is to focus some of this capability on what could be its most challenging potential application area: large-scale field construction. If U.S. robot manufacturers will not take the lead, then perhaps researchers in universities and the construction industry itself can take up the challenge. This paper is designed in part to stimulate such efforts.

Appendix. - References

1. Albus, James S., _Brains, Behavior and Robotics_, McGraw-Hill, New York, 1981.

2. Asano, Juichi, _Amphibious Bulldozer Construction Methods_, Komatsu, Ltd., Tokyo, Japan, 1978.

3. Ayres, Robert U., and Miller, Steven V., _Robotics: Applications and Social Implications_, Ballinger Publishing Co., Cambridge, MA, 1983.

4. Brady, Michael, _et al._, _Robot Motion Planning and Control_, MIT Press, Cambridge, MA, 1982.

5. Business Roundtable, _More Construction for the Money_, 200 Park Avenue, New York, NY, January, 1983.

6. Business Roundtable, _Technological Progress in the Construction Industry_, Report B-2, 200 Park Avenue, New York, NY August, 1982.

7. Duda, Richard O., and Peter E. Hart, _Pattern Classification and Scene Analysis_, John Wiley and Sons, New York, NY, 1973.

8. Engelberger, Joseph F., _Robotics in Practice_, Anacom, New York, 1980.

9. Faher, Kamilo, _Digital Communications_, Prentice-Hall, Inc., Englewood Cliffs, NJ, 1981.

10. Hunter, Ronald P., _Automated Process Control Systems: Concepts and Hardware_, Prentice-Hall, Inc., Englewood Cliffs, NJ, 1978.

11. "Japan Takes Early Lead in Robotics," _Engineering News-Record_, July 21, 1983, pp. 42, 43, 45.

12. Katz, Paul, _Digital Control Using Microprocessors_, Prentice-Hall, Englewood Cliffs, New Jersey, 1981.

13. Komatsu, Ltd., brochures on amphibious, underwater and remote-controlled bulldozer applications, Tokyo, Japan, 1978

14. Kraker, Jay, and W. Krizan (eds.), "Robots Coming to Jobsites," _Engineering News-Record_, February 10, 1983, pp. 113.

15. Kuo, Benjamin C., _Automatic Control Systems_, fourth edition, Prentice-Hall, Inc., Englewood Cliffs, NJ, 1982.

16. McDonald, A.C., and H. Lowe, _Feedback and Control Systems_, Reston, 1981.

17. "Micros Step Into Dam Grouting," _Engineering News-Record_, December 15, 1983, pp. 31.

18. Paul, Richard P., <u>Robot</u> <u>Manipulators</u>: <u>Mathematics</u>, <u>Programming</u> <u>and</u> <u>Control</u>, MIT Press, Cambridge, MA, 1982.

19. Paulson, Boyd C., Jr., "Research in the Japanese Construction Industry," <u>Journal</u> <u>of</u> <u>the</u> <u>Construction</u> <u>Division</u>, ASCE, Vol. 106, No. C01, March, 1980, pp. 1-16.

20. "Robot Makers Building-shy," <u>Engineering</u> <u>News-Record</u>, April 28, 1983, pp. 30, 32

21. Russell, Marvin, Jr., "Odex I: The First Functionoid," <u>Robotics</u> <u>Age</u>, Vol. 5, No. 5, Sept./Oct., 1983, pp. 12, 14-18.

22. Safford, Edward L., <u>Handbook</u> <u>of</u> <u>Advanced</u> <u>Robotics</u>, Tab Books, Blue Ridge Summit, PA, 1982.

23. Shanmugam, K. Sam, <u>Digital</u> <u>and</u> <u>Analog</u> <u>Communications</u> <u>Systems</u>, Wiley, New York, 1979.

24. Sinnema, William, <u>Digital,</u> <u>Analog</u> <u>and</u> <u>Data</u> <u>Communication</u>, Reston Publishing Co., Reston, VA, 1982.

25. Spectra-Physics, Inc., brochures on laser-controlled excavation and grading, Mountain View, CA, 1983.

26. Ward, Carter J., "Automatic and Adaptive Controls for Construction Equipment," Paper 750765, Society of Automotive Engineers Off-Highway Vehicle Meeting, Milwaukee, Wisconsin, September 8-11, 1975.

27. Ward, Carter J., "Hydraulic Control Designs for Retrofitting to Mobile Equipment, <u>Mobile Hydraulic Design Symposium</u>, Paper No. 43, May 13 - 15, 1980.

28. Weinstein, Martin B., <u>Android</u> <u>Design</u>, Hayden Book Co., Inc., Rochelle Park, NJ 1981.

EXPERIENCES WITH MICROS IN PROJECT CONTROLS

Dr. Gui Ponce de Leon, P.E.
Project Management Assoc., Inc.
and The University of Michigan
Associate Member, ASCE

Mr. David S. Povilus, P.E.
Project Management Assoc., Inc.
Ann Arbor, MI

Abstract: The experience of one and one-half years of microcomputer use in project controls is related, with the intent of helping the system buyer anticipate useful applications, and suggesting to the system vendor areas where new products need to be uᴄveloped.

1.0 INTRODUCTION

Common advice given to those shopping for a microcomputer is to select software first and then look at equipment that supports it. When the autnors set out to select a companywide micro-based system for project controls applications, they followed this advice closely. Because of tne general lack of micro experience in the controls field at that time, however, selecting software was a considerable problem in itselr. First, the difterences and relationships between the new system and various mini-based systems already in use had to be thought out. Then, a broad strategy for implementing micros as a practical tool for controls professionals needed to be formulated.

To the extent that suitable software has been available, the effort has been fruitful over one year of 8-bit machine usage which has now progressed to a 16-bit machine upgrade stage. It is the intent of this paper to relate our experience, as a way to point out some of the more readily achievable microcomputer uses in project controls. In addition, this experience highlights good points and problems; these, in turn, suggest criteria that may be used by the system buyer in making selections today, and by system suppliers in developing new products for the future.

In discussing microcomputer usage, it is useful to emphasize the distinction between personal versus production applications. By definition, personal usage implies a specific application where the one and only user is also the person who is best able to interpret the system output. In general, personal applications are characterized as being of short or uncertain duration, as having small amounts of data entry, and as dealing with a need or subject matter that is so closely related to the user that training of others in its utilization is not practical. In production usage, on the other hand, the amount of data is greater, and the expected life of the application will exceed some minimum. The output herᴇ is considered the product of a group and not an individual, and

20

thus there is a need for system documentation and tne training of backup users. Furthermore, different users may have different tasks within the application and some may have no familiarity with the subject matter (e.g., data entry clerks).

To date, our implementation of applications software has included word processing, database, spreadsheet, and CPM scheduling. There has also been usage of communications, BASIC language, and operating system utilities. In addition to our own micros, our staff had the opportunity to view and administer the usage of two other makes involving other operating system, database, and spreadsheet programs. Both personal and production applications in these areas will be described, and for each case where appropriate, the available alternatives to a micro will be recounted.

2.0   WORD PROCESSING

Word processing in a production mode is equivalent to replacing the secretary's typewriter with a microcomputer and suitable software. This function is more efficiently handled by a multi-user, dedicated word processing system staffed by a group of specialists; however, in some instances such a service is either not available, or cannot be counted on to meet stringent turnaround requirements, or implies unnecessary duplication of efforts.

Turnover considerations, for instance, promoted the micro for word processing at a power plant site where the time of, and required attendance at, daily meetings changed at short notice but still had to be reflected in an otherwise standard schedule cover letter. In the same group, word processing had personal usage in the production of lists of restraints reported in those meetings. In the case of some of the planners the cryptic style of their notetaking combined with their good typing skills to make learning the micro worthwhile, while for the others the site service was the better route. Incidentally, if the micro's word processing had been hard to learn, or if letter-quality print had been required, neither of these applications would have been economical.

On the other hand, at the home office a good deal of the project controls work involves writing progress reports, right-to-recovery analyses of claim issues, cost/schedule audit reports, etc. In the face of tight deadlines, our staff has found it increasingly advantageous and productive to draft the reports d. .ctly on the micro. Becoming so "hooked," has avoided time-consuming rechecking of typed text (when done by secretaries), eliminated the need for legible handwriting which is painfully slow, and minimized clerical overtime. Bypassing the typist has correspondingly reduced our budgets and quotations for work, and increased the firm's competitiveness.

3.0 DATABASE

From our experience, the most popular usage of microcomputers in project controls is with databases. On one project site, three different types of micro databases are in use, despite the fact that more powerful versions of these tools have been made available in on-line, batch, and time-share modes on larger computers. Some of the reasons for this phenomenon are legitimate, and have to do with the limited size or transitory nature of construction projects. If new information needs arise quickly due to the unexpected movement into a new phase, the lag in getting mainframe programming or equipment may be unacceptable. In this case a micro can be acquired and a database set up almost immediately, so that the capture of data can proceed in parallel with the development or modification of the mainframe system to which the data will eventually be transferred. In one instance the information needs were large enough to require a minimal micro set-up effort, but not so important or long-lasting as to economically justify involving corporate computer operations, with their stringent programming, file security, and terminal installation standards. As it may be expected, a word of caution is worthwhile in this regard. Undesirable consequences occur when a micro is chosen over an established system for political reasons. If the "owner" of an existing database is unwilling to alter it to cater to another group's coincident needs, or if a potential user is put out by the idea of following procedures and getting reports that he doesn't issue, two competing systems may result. Not only is it inefficient to enter data twice, but the confusion caused by conflicting information on project status only makes management decisions more difficult. A similar reliance on micro database software has taken place at the home office, with almost all of our work employing database structures meeting specific needs. It is clear that comparable use of time-sharing services would have been economically unfeasible.

Our own choice, a relational database, has proven invaluable both in the home office and on remote sites. Its most sophisticated use, developed for a specific client need, has been in schedule generation (ref. "Short-Term Scheduling: A Microcomputer Application", Transactions, AACE 1984 Mid-Winter Symposium).

We have also seen the use of a simple list-type micro database for keeping track of the responsibility assignment for design change package action items, for logging design changes through field engineering, and for tracking the construction status of design documents. In one of our projects, one staff has implemented a hierarchical database application which matches purchase requisitions to the scheduled activities they restrain, and statuses their trip from approval cycle through receipt inspection.

Other favored uses have related to the development of as-built data in support of cost and schedule claims analysis. One application allows the development of as-built start/finish dates for schedule activities. In this application, data comes from field dailies. Once each daily is input, together with coded key words for areas, subcontractors, systems, etc., then printouts are analyzed to identify activity patterns. Adding activity codes to the database then allows for printouts with as-built start dates, finish dates, and performance periods. From these printouts, our staff can readily complete the as-built CPM schedule.

In addition, claims analysis requires extensive reviews of project documents. Currently as these reviews take place, our staff highlights key text and cross codes the documents to type, disputed issue, preceding documents, etc. A clerk then keys this into the database, and document chronologies are easily generated.

All of the above applications are of the production type. We have seen limited personal usage of databases, even though a good part of our staff's time is spent updating, tabulating, and sorting data that he/she uses in preparing plans and schedules. At the project sites, this situation is probably due to two factors: each planner needs to use a micro only a small portion of the working week yet, with all planners demanding access according to the same reporting cycle, sharing a machine is difficult, and, secondly, most planners are too involved in an ongoing routine to spend time learning about database software.

4.0   ELECTRONIC SPREADSHEETS

Electronic spreadsheets come to mind immediately in any consideration of personal computing. Yet, in our experience of project controls, their personal usage has been limited to the tracking of intradepartmental payroll hours, and the preparation of budget and contract proposal figures. We would like to see this expanded to controls use in live contracts; in building job-specific cost estimating tables, for example. Instead, where this type of software has been employed in ongoing projects, it has taken on more of a production flavor. For example, on a power plant site, cost engineers responsible for different elements of one project provide their respective month-end totals so that the spreadsheet can be used to produce a single bulk quantities-to-go report each period. In some cases, such as for a weekly project labor summary, the spreadsheet has been used for applications that were better suited for (and were eventually transferred to) a database. Although a spreadsheet is easier to learn than a database and may save some time over manual methods, most production applications are worth developing as databases from the start.

5.0   CPM SCHEDULING

The biggest problem with the micro-driven CPM packages reviewed by the authors has been the unwillingness or inability of their vendors to explain their method of operation. This issue is important because the lack of size/reputation of many of the vendors behooves us to verify for ourselves the legitimacy of these professional tools. In addition, it is only with an understanding of the rules followed in an algorithm that a proper analysis of its outputs can be made.

Next on our list of complaints is the lack of easy access to the programs' input and output. It may not be reasonable to expect a micro CPM to generate top-notch plots, but by the same token it should be possible to transmit the data to a mainframe or mini that can. In the same vein, we would like to be able to design activity input screens tailored to each project using our own database program.

In the face of these roadblocks, we chose a software that combined a bare-bones scheduling segment with the best character-graphic charts we could find. Its experimental personal use in the field failed due to its inflexibility in activity labelling and because it couldn't integrate one planner's network with the mainframe system used for the overall project plan. A similar failure was experienced in the home office primarily due to the chosen software inability to accept progress data, such as actual starts/finishes, percent completed, actual/earned costs, etc. Currently the software is being used mostly to generate proposal schedules which are small in size and require no updating.

6.0   COMMUNICATIONS UTILITIES

Without communications utilities, it would be impossible for micros to fill the stopgap processing role described earlier under database usage. Aside from their obvious usefulness in the networking of common data between several micros, they can also extend the usefulness of an individual micro, as was the case when a large file of punchlist items exceeded the sort capacity of our micro, but could be quickly sorted by a time-shared mainframe and then returned to the micro for reporting. These utilities have also allowed the generation of input files which are then transmitted in batch mode to the mini-based timesharing service. Although this voids the on-line text editing features otherwise available, the savings are so significant that we expect to see an increase of this use of the micro.

Unfortunately, there is no guarantee that a given communications package will allow intelligent conversation with a particular mainframe without a good deal of trial and error in its configuration. Even then, establishing communication is only a first step. Because the format in which records are stored has to be consistent, one consideration in selecting a database software relates to communications: what formats will the database read from and write to.

While our usage of BASIC has been limited, we have found it important to become knowledgeable of our micro's operating system. From the onset, production usage by neophytes required the OS programming of their floppy disks to automatically set the machine

TABLE 1:   LEVEL OF SUPPORT REQUIRED FOR APPLICATIONS SOFTWARE

Software/ Application Type — Support Requirements	WORD PROCESSING		DATABASE	
	Personal	Production	Personal	Production
APPLICATION DEVELOPMENT	low	low	low	high
INTEGRATION WITH OTHER APPLICATIONS	low	medium	low	possibly high
EDUCATION OF USERS	low	high	high	low
OPERATING SUPPORT	low	low	medium	medium

Software/ Application Type — Support Requirements	SPREADSHEET		CPM SCHEDULING	
	Personal	Production	Personal	Production
APPLICATION DEVELOPMENT	low	medium	high	high
INTEGRATION WITH OTHER APPLICATIONS	low	possibly medium	high	high
EDUCATION OF USERS	medium	low	medium	high
OPERATING SUPPORT	low	high	high	high

TABLE 2: LEVEL OF SUPPORT REQUIRED FOR UTILITIES SOFTWARE

Software/ Application Type Support Requirements	COMMUNICATIONS		BASIC LANGUAGE	
	Personal	Production	Personal	Production
APPLICATION DEVELOPMENT	medium	high	medium	high
INTEGRATION WITH OTHER APPLICATIONS	high	high	n/a	n/a
EDUCATION OF USERS	high	low	high	low
OPERATING SUPPORT	medium	medium	medium	high

Software/ Application Type Support Requirements	OPERATING SYSTEM	
	Personal	Production
APPLICATION DEVELOPMENT	low	high
INTEGRATION WITH OTHER APPLICATIONS	n/a	n/a
EDUCATION OF USERS	high	low
OPERATING SUPPORT	low	medium

up for their individual applications. When we went from
floppy-only machines to a type with a hard disk, it was evident
that the shared usage of any one machine by several production
applications resulted in file management problems. These were
solved by instituting a new set of procedures and by programming a
special set of function keys for each installation to control each
user's access to the hard disk upon booting-up.

7.0   IN CONCLUSION

In planning the intended use of a microcomputer the amount of
support the end user(s) will need should be taken into
consideration. Tables 1 and 2 indicate the levels of support
demanded by the applications described above. Within our company,
support is provided by a hierarchy of knowledgeable people. Each
software product has been taken on by one of our professionals for
thorough study and testing. Each installation is assigned to a
machine operator who is the first line of defense in answering
routine user questions. All questions beyond the knowledge of the
operators are referred to a single expert who either solves the
problem or interfaces with either our own software specialists or
the proper vendor for a solution. This expert currently spends one
day per week in support of seven micros.

The most significant statement that the authors can make from
their experience to date is that there is a dearth of adequate CPM
scheduling software for microcomputers. A recent survey of
computer users in construction by Wilson, Ma, and Associates, Inc.,
confirms this conclusion, with planning and network scheduling
cited as the second most needed application. Our definition of
adequacy would set a high standard. Included in our specifications
would be full on-line text editing, extensive handling of progress
conditions, a level of character graphics capabilities, resources
and cost integration, and analytical features comparable to that
available on mainframes and minis; however, the system can be
designed for personal usage, say, 500 to 1000 activities. This
program would be a tool for a planner to manage in detail a small
project or that portion of a large one for which he is responsible.
For this reason both input and output should be accessible in terms
of the major micro databases on the market. In addition, the
software should be written in such a way that its output, or a
summary of its output can be sent to any of the popular high-end
project controls systems for plotting and inclusion in overall
project network calculations and reports.

MICROCOMPUTERS FOR CONSTRUCTION FIELD OFFICES

Michael J. O'Connor*, Carl E. DeLong*,M.ASCE,
and Glenn E. Colwell*

ABSTRACT

This paper presents an overview of U.S. Army Corps of Engineers experience with using microcomputers for construction management. Construction management applications for contract administration and office management, typical hardware and software, and field office implementation are discussed.

INTRODUCTION

The Corps of Engineers, within its world-wide Military Construction and Civil Works mission, manages the placement of several billion dollars worth of construction each year. Existing microcomputer technology provides the Corps the opportunity to acquire cost effective computer assistance at its construction field offices.

The objective of using microcomputers at field offices is to enhance the Corps' ability to construct quality projects on time and within budgets. Microcomputers improve construction management effectiveness by supporting more informed decisions.

CONSTRUCTION MANAGEMENT APPLICATIONS

Corps Resident Engineers are using microcomputers for several construction contract administration and office management functions.

Typical contract administration uses include: construction project management via Critical Path Method (CPM) network analysis, general contract information, change order, and submittal management.

Commercially available project management systems are being used to 1) review and evaluate contractor submitted construction schedules, 2) monitor actual versus scheduled

*   US Army Construction Engineering Research Laboratory, P. O Box 4005, Champaign, IL 61820

progress, 3) estimate partial payments based on work placed, 4) evaluate the impact of change orders, 5) assess contractor claims, and 6) perform "what-if" analyses to ensure that projects are completed on time and within budget.

The ability to perform these analyses at the resident office represents a significant increase in Corps management capability. The Corps has traditionally required its contractors to use CPM network analysis to schedule and monitor job progress. However, its effectiveness has been limited by lack of direct access to computerized computational support. In general the contractor's job superintendent and Corps field managers have had to rely on home office or third party construction management consultant support for network analysis. This arrangement was too slow; it provided little more than historic information. CPM's that don't reflect current actual job conditions don't support informed management decisions. Resident office microcomputers provide the capability to use network analysis effectively.

The Corps is also using data base management systems (DBMS) to develop Corps-specific construction management applications. General contract and change order status reports are shown in Figures 1 and 2. Change orders are tracked from the date the Corps requests a proposal from the contractor through the date of formal contract modification. Revisions to the original contract price and contract completion date, based on finalized contract modifications, are automatically reflected in the contract information report.

Corps contracts typically require various contractor submittals such as shop drawings, material samples, certificates of compliance, etc. Figure 3 is a submittal register report showing the status of contractor submittals. The submittal register program provides the Resident Engineer two exception reports; delinquent submittals and overdue approvals. Delinquent submittals are those which have not been submitted by the contractor's scheduled date. Overdue approvals are those submittals which have been received by the Corps for which approval action has not been completed within the time allowed by the contract. In addition to tracking actions that have slipped, both of these reports can be used to look ahead or backward from the present. By asking for these exception reports as of some future date the Resident Engineer can plan for future submittal review workload. Looking back he can reconstruct submittal status for any past date.

Microcomputers are also being used for various field office management functions. Figure 4 presents a spread sheet example for managing work placement and supervision

CONTRACT INFORMATION REPORT

RESIDENCY:gecdemo

DATE: 05DEC83

IB/Contract No: DACA88-83-C-4321   Project Description: sewage treatment plant - vafb
Contractor: j.c.nickel &sons
Cost: $5125000

Plans & Specifications Complete: 15jul82	Constructibility Review: 01jul82	Plan in Hand Survey: 20jul82	Comments Submitted: 25jul82	Bid Opening Date Scheduled: 15aug82	Bid Opening Date Actual: 12sep82
Contract Award Date: 24sep82	Preconstruction Conference 30sep82	Environmental Plan: 28sep82	Quality Control Plan: 04oct82	Submittal Register 15oct82	Safety Plan: 15oct82
Construction Schedule: 10oct82	Notice To Proceed: 05oct82	Progress Scheduled: 50%	Progress Actual: 48%	Original Completion Date: 15nov84	Lost Time Accident: 0
Photographs: 50%	Modifications Total Issued: 7	Modifications Total Open: 3	Modifications 90-Days Old: 2	Modifications 2-Part Open: 2	Disputes Pending: 2
Revised Contract Price: $ 5491500	Revised Contract Completion Date: 24JAN85	Last Payment for:NOV83 Amount:$ 50000	Number of Mods Executed: NOV83 No: 1	Final Acceptance Scheduled:	Final Acceptance Actual:
DD Form 1354 Issued: Returned:	Shop Drawings Transmitted:	O&M Manuals Transmitted:	As-Built Drawings Completed:	GFP Final Inventory:	Value of Mods Executed: NOV83 Value: $ 50000
Letter Clearing Deficiencies:	Architect/Engineer Evaluation:	Contractor Performance Evaluation:	Claims Pending: 0	Final Payment Processed:	Files Retired:

Post Completion Inspections:
        4-Month   9-Month   Remarks:
Scheduled: 15mar85   15aug85
Actual:
Reported:

Figure 1

MODIFICATION   STATUS   REPORT

RESIDENCY: gecdemo   AS OF: 05DEC83

                                                          ORIGINAL CONTRACT PRICE: $5125000
CONTRACT NO: DACA88-83-C-4321   PROJECT: sewage treatment plant - vafb

                                                          PRESENT CONTRACT PRICE: $ 5491500

                            NOTICE TO PROCEED: 05oct82   ORIGINAL COMPLETION DATE: 15nov84

CONTRACTOR: j.c.nickel &sons   REVISED COMPLETION DATE: 24JAN85

MOD	DESCRIPTION	DATE TO CONTRACTOR FOR PRICE/ PROP.DUE	DATE PROPOSAL RECEIVED/ PRICE/TIME	DATE NTP ON 2-PART	DATE CONTRACTOR SIGNED	DATE FINAL PART IS EXEC.	TIME ALLOWED	FINAL PRICE
0001	revise manhole £27	15NOV82 25NOV82	25NOV82 $22000 7 DAYS		15DEC82	15DEC82	5 DAYS	$21500
						Remarks : EXAMPLE OF 1-PART MOD FINALIZED		
0002	increase size of influent conduit from 48" to 72"	12NOV82 01DEC82	11DEC82 $600000 90 DAYS	05MAR83			0 DAYS	$
						Remarks : EXAMPLE OF 2-PART MOD STILL BEING NEGOTIATED		
0003	changed site condition - rock under digester at el. 765.5	25MAY83 05JUN83	01JUN83 $735000 35 DAYS	25MAY83	02JUN83	03JUN83	15 DAYS	$225000
						Remarks : EXAMPLE OF 2-PART MOD FINALIZED(SEE DISPUTE 2)		
0004	increase size of visitor parking lot	25JUL83 25AUG83	25SEP83 $73500 20 DAYS		20OCT83	13OCT83	20 DAYS	$70000
						Remarks : EXAMPLE OF 1-PART MOD FINALIZED		
0005	add new 5000 gpm lift station	15JUL83 15AUG83	12AUG83 $ 15 DAYS	15JUL83			0 DAYS	$
						Remarks : EXAMPLE OF 2-PART MOD STILL NEGOTIATING		
0006	repair erosion from unusually heavy rains (20-25SEP83)	30SEP83 15OCT83	20OCT83 $250000 180 DAYS		15NOV83	17NOV83	30 DAYS	$50000
						Remarks : EXAMPLE OF 1-PART MOD FINALIZED		
0007	add sodding to berms of digester and trickling filter	01NOV83 20NOV83	$ 0 DAYS				0 DAYS	$
						Remarks : EXAPLE OF 1-PART MOD WAITING ON CONTR PROPOSAL		

DISPUTES: ID   DESCRIPTION   DISPOSITION

        PENDING: 0001 differing site conditions awaiting further justific
                       -water table   ation from contractor

        SETTLED 0002 subsurface conditions -   mod£0003 issued-contracto
                       rock encountered   r's claim valid

Figure 2

```
                        SUBMITTAL REGISTER

RESIDENCY: DEMO1                    TODAY'S DATE: 01MAY83
CONTRACT NUMBER: DACA88-83-C-0123

   1        2          3           4            5         6      7     8
SUBMITTAL  SPEC PARA  DESCRIPTION  CONTRACTOR   CORPS     REVIEW ACT   REM
& ITEM     NUMBER(S)  OF MATERIAL  SCH DATES    ACTION DATES  BY  ION   ARK
---------  ---------  -----------  ----------   ------------  ------  --------

0001-1     4A,2.1     CT CONCRETE  SS 01FEB83   AS 15MAR83   AREA   B     Y
I=                    MASONRY      AN 20FEB83   AA 01APR83
J=                    UNITS        MN 01APR83

           REMARKS: Certificate outdated - to be replaced.


0001-2     4A,2.11    CT ANCHORS,  SS 01FEB83   AS 15MAR83   AREA   C     N
I=                    TIES,JT REI  AN 20FEB83   AA 01APR83
J=                    NFORCING     MN 01APR83

0001-3     4A,2.2     SS CMU,TIES  SS 20MAR83   AS 15MAR83   AREA   A     N
I=                    JT REIN, &   AN 10APR83   AA 01APR83
J=                    ANCHORS      MN 10APR83

           7D,3.1     CT TT        SS 01APR83   AS           AREA         N
I=                    CAULKING     AN 15APR83   AA
J=                    COMPOUND     MN 30APR83

006-1      7D,3.3     SS           SS 20APR83   AS           AREA         N
I=                    CAULKING     AN 05MAY83   AA
J=                    COMPOUND     MN 05MAY83

0002-1     8A,4.2     BLDRS HARD-  SS 15MAY83   AS 01APR83   AE     A     N
I=                    WARE,APPROV  AN 01JUN83   AA 24APR83
J=                    ALS LIST     MN 05JUN83

0002-2     8A,4.3     HARDWARE     SS 15MAY83   AS 01APR83   AE     A     N
I=                    SCHEDULE     AN 01JUN83   AA 24APR83
J=                                 MN 05JUN83

0003-1     8C,3.1     SD METAL     SS 15FEB83   AS 10APR83   AE     A     N
I=                    DOOR FRAMES  AN 01MAR83   AA 24APR83
J=                                 MN 01MAR83
```

Figure 3

** MILITARY **

WORK PLACEMENT AND S&A SCHEDULE

(IN THOUSANDS)

			OCT	NOV	DEC	JAN	FEB	MAR	APR	MAY	JUN	JUL	AUG	SEP
MONTEREY	WPE	B								9033	11447	13210	16127	19200
		A								9313				
	S&I	B	28	98	154	225	281	337	393	463	519	576	646	730
		A								460				
	S&I RATE	B	0	0	0	0	0	0	0	5.1	4.5	4.4	4	3.8
		A	0	0	0	0	0	0	0	4.9	0	0	0	0
SACRAMENTO	WPE	B								17633	20945	24409	27407	29872
		A								15718				
	S&I	B	30	104	163	237	296	355	415	489	548	607	681	770
		A								491				
	S&I RATE	B	0	0	0	0	0	0	0	2.8	2.6	2.5	2.5	2.6
		A	0	0	0	0	0	0	0	3.1	0	0	0	0
UTAH	WPE	B								3286	3889	4816	5575	6141
		A								2665				
	S&I	B	13	45	71	103	129	155	180	213	238	264	296	335
		A								217				
	S&I RATE	B	0	0	0	0	0	0	0	6.5	6.1	5.5	5.3	5.5
		A	0	0	0	0	0	0	0	8.1	0	0	0	0
VALLEY	WPE	B								441	653	974	1107	1242
		A								242				
	S&I	B	1	3	4	6	8	9	11	13	14	16	18	20
		A								16				
	S&I RATE	B	0	0	0	0	0	0	0	2.9	2.2	1.6	1.6	1.6
		A	0	0	0	0	0	0	0	6.6	0	0	0	0
OTHER	WPE	B												
		A												
RES OFF TOTAL	WPE	B	0	0	0	0	0	0	0	30393	36934	43409	50216	56455
		A	0	0	0	0	0	0	0	27938	0	0	0	0
	S&I	B	71	250	392	571	713	856	999	1177	1320	1463	1641	1855
		A	0	0	0	0	0	0	0	1184	0	0	0	0
	S&I RATE	B	0	0	0	0	0	0	0	3.9	3.6	3.4	3.3	3.3
		A	0	0	0	0	0	0	0	4.2	0	0	0	0
DISTRICT OFC	S&I	B	45	158	249	362	452	542	633	746	836	926	1039	1175
		A								849				
	S&I RATE	B	0	0	0	0	0	0	0	2.5	2.3	2.1	2.1	2.1
		A	0	0	0	0	0	0	0	3	0	0	0	0
TOTAL S&I	S&I	B	117	408	641	932	1165	1398	1632	1923	2156	2389	2680	3030
		A								2033				
	S&I RATE	B	0	0	0	0	0	0	0	6.3	5.8	5.5	5.3	5.4
		A	0	0	0	0	0	0	0	7.3	0	0	0	0
OVERHEAD		B	19	67	106	154	192	231	269	317	356	394	442	500
		A								313				
	RATE	B	0	0	0	0	0	0	0	1	1	.9	.9	.9
		A	0	0	0	0	0	0	0	1.1	0	0	0	0
TOTAL S&A		B	136	475	747	1086	1358	1629	1901	2240	2512	2783	3123	3530
		A	0	0	0	0	0	0	0	2346	0	0	0	0
	RATE	B	0	0	0	0	0	0	0	7.3	6.8	6.4	6.2	6.3
		A	0	0	0	0	0	0	0	8.4	0	0	0	0

Figure 4

and administration (S&A) costs/rates. In this example monthly estimated work placement (WPE B) and supervision and inspection (S&I B) are forecast for each resident office and the district. Total S&A costs are calculated by adding district overhead costs to total S&I costs. Forecast S&I and S&A rates are calculated by dividing costs by work placement. Actual work placement (WPE A), S&I (A), and S&A(A) are posted throughout the year and the corresponding S&I/S&A rates are calculated for comparison to forecasted rates.

Other office management applications include Resident Office budgeting, funds control, property accounting, and personnel vacation and training schedules.

This type of "look-ahead" capability provided by the contract administration and office management applications supports pro-active versus re-active management. It doesn't take a bad manager to make bad decisions; incomplete or wrong information can sabotage the most conscientious manager. The Corps microcomputer based decision support systems help provide the right information at the right time.

MICROCOMPUTER SYSTEMS

The Corps has not standardized on a single microcomputer for all field offices. The rapid rate of microcomputer technology change and the decentralized and diversified nature of the Corps' mission makes standardization impractical and undesirable. Attempted standardization, in the face of rapidly changing technology, would not only ensure almost immediate obsolescence, but also perpetuate obsolescence throughout the life of the standard. Necessary and sufficient compatibility between micros, and among micros and mini/main frames, is provided by standard communication protocols and de facto microcomputer industry standards.

Hardware

The success of the IBM-PC, has established the 16-bit microcomputer in the marketplace. The increased random access memory (RAM) capacity of the 16-bit processor represents a significant improvement over 8-bit processors.

While the Corps currently has several microcomputers in field office use, including 8-bit microprocessors, a typical microcomputer purchased for field office use at this time would have at least a 16-bit microprocessor (an 8-bit co-processor would be optional), a minimum of 256 KByte RAM, a 10 MegaByte Hard Disk, a 5.25 inch 320-400 KByte floppy disk, and a bus structure with open slots for various extension boards. A monochrome monitor with advanced video features, such as 80/132 columns and reverse video, and a keyboard with programmable function keys would be included. The printer would be of good correspondence quality with both friction and tractor feed and capable of printing 132 character lines on 14 inch paper. Communications would be provided by an asynchronous auto-dial/auto-answer modem using the Bell standards at 300/1200 baud.

The operating system, would include MS-DOS/PC-DOS and CP/M-86. (CP/M-80 for systems with 8-bit co-processors would be optional.)

Software

Typical software at this time would include a project management system, a data base management system, a spread sheet, a word processor, and a communications program.

The project management system would be capable of processing networks of 2000 to 3000 activities in either activity-on-the-arrow or precedence notation.

Currently, dBaseII is the de facto standard DBMS that would be typically used. Several utilities have been developed for dBaseII which enhances its capability.

VisiCalc, SuperCalc, and Multiplan are typical of the type of required spread sheets. The trend towards integrated graphics, exemplified by SuperCalc III, increases the utility of spread sheet programs.

WordStar, or equivalent, would be included for word processing.

Several communications packages are commercially available. However, they are limited by unique protocols. Transfer of files between microcomputers with error checking requires that the same communication program be on both computers. Commercial communications programs seldom run on several different microcomputers. However, a public domain program, modem7, does run on many different microcomputers. In addition, modem7 compatible programs have been written for other operating systems such as UNIX. Modem7 is available from CP/M User Groups and many of the dial-in remote CP/M systems. The MS-DOS Users group has also developed a 16-bit version of modem7. An additional benefit of modem7 is that since it is in the public domain users can modify the source code as required.

Future Trends

     We expect to see an active marketplace characterized
by continued technology advances. The 32-bit desktop
microcomputer is on its way. New operating systems will
emerge and UNIX will become more available on low cost
systems, especially multiuser systems. Operating systems
allowing concurrent operations for single users will become
more common. However, the 16-bit microcomputer will be the
mainstay for the near future.

     We also expect that more powerful and easier to use
data base management systems will be developed specifically
for 16-bit machines. Second generation integrated software
combining true data base management systems, spread sheets,
graphics, word processing, and communications will emerge
in the near future.

IMPLEMENTATION

     The U.S. Army Construction Engineering Research
Laboratory (CERL) is supporting the implementation and use
of microcomputers in construction field offices through
several Corps-wide initiatives. The successful transition
from manual procedures to the use of microcomputer systems
depends upon knowledgeable users and properly operating
hardware and software. Training and hardware/software
maintenance are essential and the cost is well justified.

Corps-wide Support

     CERL is the Corps' principal laboratory for
construction management microcomputer use and as such
serves as the focal point for Corps-wide information
exchange. CERL has published the "Microcomputer Selection
Guide for Construction Field Offices"(1) which presents
basic guidance for Corps field offices interested in
acquiring and using microcomputers. The Guide includes an
introductory tutorial on microcomputer hardware and
software, a discussion of construction field office
functional uses of microcomputers, procedures for
performing a functional analysis, a microcomputer selection
procedure, and guidance on system approval and procurement.
The Guide is currently being revised. The updated Guide
will be published in August 1984.

     The "Construction Micro Notes"(2) newsletter,
published three times a year, contains articles pertaining
to field office experiences, recommendations, and technical
data for those using or planning to use microcomputer
systems. Newsletter articles are authored by Corps field
offices and CERL.

     The Construction Microcomputer Users Group (CMUG) has

been established to allow users to directly exchange ideas and lessons learned, discuss existing and future applications, and establish short and long range Corps goals. The CMUG, which meets semiannually, conducted its third meeting in Atlanta, GA on 2-3 May 1984.

The Corps' Construction Applications Library is used to exchange Corps developed or public domain application programs among microcomputer users. Table 1 lists the applications currently in the library. The library is maintained on-line so field offices can electronically access program abstracts to determine program applicability. Programs can also be electronically transferred to microcomputer users.

Table 1

CONSTRUCTION APPLICATIONS LIBRARY

Application	Author*	Required Software**		
		dB	B/WS/CS	SC
Submittal Register	CERL	x		
Modification Status	CERL/NCS	x	x	
Pay Estimate	NCS/SPK	x	x	
Contract Information	CERL	x		
Forecasted Work Placement	SPK			x
Work Placement, S & A	SPK			x
Res. Off. Operating Budget	SPK			x
Cost and Earnings Report	SPK			x
Emergency Operations	SPK	x		
Funds Status	NCS		x	
Construction Progress	NCS		x	
Construction Status	NCS		x	

```
    *    CERL: Construction Engineering Research Laboratory
         NCS:  St. Paul District
         SPK:  Sacramento District

   **    dB:   dBASEII          B :  BASIC
         WS:   WordStar         CS:  CalcStar
         SC:   SuperCalc
```

CERL has also developed a computer-resident microcomputer knowledge base which contains information on hardware, software, system selection, Corps-specific application programs , and Corps policy. This system allows Corps microcomputer users to electronically access the knowledge base, input their comments, ask questions, and receive answers from designated content experts.

Training

Corps training is obtained from hardware/software vendors, from Corps District ADP personnel, from previously trained field office personnel, and to some extent from participation in microcomputer users groups, and manuals.

Those individuals who are to be the primary users of the system receive initial training prior to or concurrent with delivery of the system. This training provides an understanding of the overall system, the physical interconnections, and how to access and use the major software packages. This initial training is most important and must be done by a competent instructor, using equipment and programs identical with those to be used by the field offices. This training provides liberal hands-on practice and clearly written handouts for reference.

Within the first six months of operation those who received the initial training are provided advanced training to cover system options and capabilities not previously addressed. This phase of training should be presented using the same format as the initial training.

Maintenance

Commercial organizations provide a variety of maintenance service options for microcomputers and peripheral units. The cost is highest when service is to be provided on short notice at the field site; and least when the equipment to be serviced is delivered to the vendors shop. All field office microcomputer equipment is covered by a hardware maintenance contract. Generally, these contracts provide for service at the vendors shop with a turnaround time not to exceed 72 hours. The need for faster response and/or on-site service depends on local circumstances.

In general, 'software maintenance' is the term used to describe an agreement wherein the purchaser of a software package receives subsequent revisions, improvements, etc. from the software package developer. Some software contains the cost of such service in the initial purchase price, while other software packages require an additional annual fee for software maintenance. Software maintenance is desirable and consequently is obtained in all cases where the cost is reasonable.

CONCLUSIONS

Corps of Engineers Construction field offices have many applications that benefit from the use of microcomputers. The Corps has found microcomputers to be cost effective not only from the standpoint of improved productivity, but also from the enhanced quality of decisions possible when better, more complete and timely information is made available to the Corps managers. Better management decisions result in decreased construction costs through a reduction of government caused delays, better change order impact analyses, and improved claims management.

REFERENCES

1.   Grobler, F., O'Connor, M. J., Colwell, G.E., "Microcomputer Selection Guide For Construction Field Offices," USA CERL Technical Report P-146, June 1983.

2.   "Construction Micro Notes" Information Exchange Bulletin, USA CERL, Published three times a year.

SMALL COMPUTERS IN CONSTRUCTION: A CASE HISTORY
MORSE/DIESEL, INC.

MICHAEL BOWERMAN*

I. INTRODUCTION

Our continuously changing modern society has dictated that
today's developers and builders respond in new ways to meet new
challenges. Economic swings require quick reponse to meet market
conditions. Ever increasing land and property values continue to
increase the total cost of projects. Inflation and escalation
fluctuations require constant monitoring and forecasting.
Interest rates, which have seen record highs, have forced
unprecidented complexity in financing schemes, adding to the
project team a host of new participants called partners, managing
partners, limited partners, construction lenders, and mortgage
lenders all of whom want to be kept equally appraised as to
project status, finances, schedule and their respective financial
exposures.

Increasing subcontractor sophistication and, perhaps as an
outgrowth of the "litigious society" disease, a spiraling growth
in the number of financial claims against projects has created
the single largest new risk to the success of building projects
today. Extra work claims, escalation claims, coordination
claims, delay claims, and compression claims all occur with
increasing regularity.

It is this new environment which calls builders to respond with
new technology and with innovation. Our methods and our tools
must improve to meet the challenge

---

*M. ASCE; Assistant Vice President, Senior Project Manager,
Morse/Diesel, Inc., Boston, MA

The building industry, today, has responded with new heights of achievement. New tools and technologies in building construction are making events that used to be labeled 'amazing', now commonplace. We are pumping concrete higher than 1000 feet in the air. Modern cranes and lifing equipment enable us to lift loads heavier than ever before and to put them in places we couldn't previously reach. We now look to laserbeams to guide us in tunneling, to direct earthmoving and scraping equipment, to keep our buildings straight and level, and to aid in all forms of surveying. Mass production of components as larger units is improving productivity. Advances in design and material sciences are allowing us to achieve stronger and more durable structures.

Project Management is also responding with increased sophistication. Senior officers of many builders are drawn with increasing regularity from disciplines outside the building field. Financial, economic, and legal training is now supplementing the traditional technical engineering and construction curriculum. Young engineers now coming out of school are as comfortable and as adept with computers as we once were with the slide rule of past eras. Not only are they adept with the computer, but their expectations are that today's modern builders are taking full advantage of these tools which are so commonplace in academics.

So, to respond to the challenges of the marketplace and it's demand for innovation, and to maintain an attraction for the young engineers being trained today, the building industry must learn to utilize the tools of today's management technology. One of these tools is the micro computer.

This paper will evaluate the experiences of one particular builder during it's research, evaluation, and implementation of the micro computer.

II. COMPUTER COMMITTEE -- RESEARCH

Throughout the late 1970's, Morse/Diesel managers recognized that with the development of new tools and technologies in management, higher levels of control could be maintained over construction projects. Scheduling, estimating, budgeting, forecasting, and cost control functions were becoming increasingly complex and, given the high interest rates of the early 1980's, timing and control of projects was essential for their success. The publicity that the micro computer revolution was receiving, the achievement of reasonable price levels for computer hard and software, and the development of easy to use programs provided the

the opportunity for senior management to raise the question of computer applicability to project control.

A quarterly meeting of the Corporate Operations Committee created a formalized Computer Committee charged to investigate computerized estimating, scheduling and other management tools, and, the enchancement of the corporate computerized mainframe accounting and management systems. The committee was chaired by a senior vice president and was composed of officers involved with the estimating function, accounting and cost control, consulting, and operations. The membership was subsequently expanded to include the manager of the data processing department and a senior project manager with special interest in computers.

The members began researching computers in construction and recent developments of applications in each of the areas of interest. The material available was overwhelming. Magazine articles, mail order brochures, books, personal contacts, seminars, consultants, and salespeople were available in great quantity. At times, the amount of information seemed to confuse the issues and make it difficult to see which products and approaches were appropriate for the company's needs. But a continued exchange of information and communication between the committee members aided their work.

Since the directive from the Operations Committee had not specified what type of computer systems to investigate, or what approach to take, the Computer Committee members did not limit themselves to any particular type of system or approach, initially. Mainframe applications were investigated by the manager of data processing. Outside scheduling consultants and services were interviewed for possible involvement. Estimating services linked by terminal were evaluated. Micro and mini computers were examined for size, price, speed, compatibility with the mainframe, and the availability of software. The evaluation of this material was time consuming and since all members had operational duties, the first formalized meeting of the Computer Committee was four months after its creation.

This meeting, as most organization meetings go, produced few tangible results. The time was spent reviewing what various members had discovered in their respective research and specifically trying to respond to questions raised by the company president in a memo to the committee chairman. Discussions ranged from scraping the mainframe to relying solely on consultants for computer applications. Special interest groups also surfaced. The V.P. from estimating was convinced that computers had no application in the work of his department; the

V.P. from accounting wanted total control over any computing
functions so as to protect confidential corporate cost
information; the V.P. from Chicago wanted to implement
state-of-the-art micro-computers immediately because of staff
pressure. The officers were also divided between the older
senior managers who were not yet comfortable with computers, and
the younger officers who had grown up in the computer age and had
had the benefit of a computer education. The senior officers
controlled the company's financial resources and their support
was required if the project was to succeed. Establishing a
strategy to accomplish the desired goals and satisfy the
different interest groups would prove to occupy much of the
discussion time.

The meeting closed with a few distinct directions: 1. The
manager of data processing, a new addition to the corporate
staff, was convinced that the mainframe software could be updated
adequately to provide the basis for the management information
system senior managers desired, 2. He also felt that a
reasonable cost micro computer could be found to interface with
the mainframe and take the place of the currently used 'dumb'
remote terminals, and 3. The Chicago office would use on an
experimental basis a scheduling service bereau for schedule
preparation. The members parted to return to their respective
cities and offices and to continue their research and
investigations.

Over the next two months the committee or parts of the committee
met five more times in preparation for a presentation of its work
to the entire management staff at the company's yearly management
conference. Members of the committee met in Boston at a large
project jobsite to evaluate the potential uses of computerized
reporting and analysis and to continue to work on a long-term
strategy for the utilization of electronic computing within the
company. Several members visited the facilities of a schedule
consulting company to evaluate that approach. A vendor made a
demonstration of graphic plotters in the corporate offices and a
'for-lease' estimating service gave demonstrations both in New
York and Chicago.

Throughout this investigation period the committee tried to
target the potential uses for computers in the course of
Morse/Diesel's business. Word processing and spread sheet
analysis were the most obvious. Logging of documents, shop
drawing control, inventory control, budget comparisons against
actuals, forecasts of expenditures, graphic representation of
expenditures and forecasts, and standardized reporting formats
were tasks that could assist in the monitoring of project data.
These procedures would be performed independently of the mainframe

to ease the burden of home office processing. But, the transfer of pertinent summary data could be accomplished with the same machine and data would only need to be entered once. The committee found it easy to identify potential applications for the machines, but needed to convince the body of managers and convey the enthusiasm for computers that was building within the committee itself.

The Computer Committee presentation at the management conference included a demonstration of a micro computer which the data processing manager had procured. It was compatible with the mainframe and performed all the analysis that the members of the committee had identified as beneficial to the company. In the presentation several committee members also discussed the status of the trials of the scheduling service bureau and the estimating system which was being used by the Chicago office.

The report was enthusiastically received. Many managers present voiced agreement that they had a need for the enhanced analysis and record keeping capabilities that the micro computers would provide. Although there were still skeptics present, the consensus was to proceed with procurement of several test machines and continue the evaluation of the scheduling and estimating systems.

III. COMPUTER COMMITTEE -- IMPLEMENTATION

Micro computers were ordered for four test sites: Boston, Chicago, New York (controller), New York (data processing). Each office began an education and experimentation phase to familiarize themselves with the operation of the hardware, the use of the software, and the jargon of the computer era. The first step was to read the manual which was supplied with the machines, and to understand the characteristics of the 'operating system', CP/M (which stands for Control Program/Microprocessor). The operating system organizes the various functions of the micro computer like a conductor directs an orchestra, calling on different parts as needed. It is necessary to understand how this system works in order to be able to use the various functions of the machine and its programs.

Morse/Diesel personnel had little difficulty in mastering the basics of machine start-up, although, all experienced minor challenges with some phase or another. Probably most often experienced was a problem setting up the switches between the computer and the printer. If compatible signals aren't sent and received, gibberish is printed. After some trial and error, and further reading in the manuals these problems were overcome.

Initial system configuration consisted of a micro computer with two disk drives, a letter quality printer, a modem for telephone communication, and software. The start up software package was word processing by WordStar, spreadsheet by SuperCalc, and data base system by dBase II. The cost of this total package was under $9,000.

The Boston users began immediately taking advantage of the word processing program. Three secretaries learned to use the program very quickly. It aided them in the heavy correspondence load for which they were responsible. It should be noted, however, that some people are more prone to pick up the techniques than others, as was the case with Boston secretaries, but none found that they couldn't learn the program. One used the computer so much that her time had to be restricted so she could do other work and others could also use the machine. Other users in the Boston office were the cost accountant, project managers and assistant project managers, even the field superintendents showed interest in using the spreadsheet program to prepare checklists, punchlists, schedules, and reports. The readers of the superintendents' reports and lists were also delighted since the report was now decipherable without their handscript. Boston also began work on a shop drawing logging and tracking program using the data base management system. The program would keep track of submittals and their anticipated return from approval. The project manager could then review a current status at any point in time with the architects and the owners in an attempt to avoid appoval delays. This program underwent many modifications and revisions, and is presently in use in several offices.

In Chicago, the machine was primarily used for cost accounting problems. A client was using a micro computer for billing purposes. Morse/Diesel's micro was used to interface with theirs. Chicago's project managers also developed cost forecasting models for various types of hospitals; combining variable costs, locations, and space usage they were able to show the client alternative programs and their respective cost analysis'. The micro computer also functioned as a data entry teminal for the trial estimating and scheduling service bureaus and for the corporate cost accounting system.

The New York computer activity produced many standardized reporting formats. Of particular usefulness are the Equal Employment Opportunity monitoring reports for various Local, State and Federal authorities. Also, cash flow reporting and projections of cash flows as influenced by anticipated escalations and wage increases. Timing of proposed projects can be asseseed in this fashion.

The manager of data processing in New York served as the overseer of the developmental stage. He would work with the vendors on supply and distribution problems, assist in the installation of machines, run down answers to questions raised by the users, and distribute the information developed by the various experimenters. He attended a clinic for users of the data system and was able to help with dBase problems. He met with suppliers of other peripheral equipment to evaluate their merits. However, his primary duty was the management of the mainframe computer so micro computer development often had to take a back seat. But as the users increased in number and in confidence, they were able to run down their own answers. It is important, though, that the manager of data processing still be kept informed and in charge of the expansion of the systems.

Morse/Diesel expanded the number of the micro computer systems. Other programs were evaluated and purchased. Multi-Plan, a spreadsheet and reporting program, was purchased and is preferred by some users. Critical path scheduling programs were evaluated. There are many on the market, but most have limitations. One program the company is currently using is easy to use, prints a coherent bar chart output, but is difficult to revise. Other scheduling systems are being evaluated. Also, software developers are continuing to introduce new systems to the market and the old ones are continually being upgraded.

IV. FUTURE PLANS

The acceptance and utilization of the micro computer by those involved in the trial program seems to have secured its place as a standard project management tool at Morse/Diesel. As new projects are undertaken, each is fitted with one or more computers instead of the old data input terminals. The manager's ability to analyze, keep records, improve communicatios, maintain correspondence, and defend against unfounded claims is increasing as the user begins to understand the power of the computer.

The Computer Committee will remain active in overseeing the use of computers within the company. It will deseminate information to the various offices and users, make available the resources of the available company talent, and serve as a clearing house for standardization. Individual user creativity still guides specific application development, but the Committee will distribute examples for specific adaptation and to serve as catalysts for other ideas.

It is recognized that the computer brings nothing new to the
problem solving environment, nor will it organize the
undisciplined, however, the computer provides the means to
magnify the managers own ability. The value of the management
expertise Morse/Diesel brings to a client is being enhanced by
the utilization of the computer as a tool for record keeping,
analysis, and the identification of alternative courses of action.

Builders must continually look for ways to respond to the
changing needs of the marketplace. Only through investigative
experiences like those described here, can the industry hope to
meet the challenges in today's building environment.
Productivity enhancement must begin in the front office. We must
continue to challenge ourselves and one another to find new ways
to keep abreast of technology, implement modern techniques, and
increase the value of our services to consumers of construction
expertise.

APPENDIX: MORSE/DIESEL, INC. BACKGROUND

Morse/Diesel, Inc. is a construction management firm which provides consulting, management and supervisory functions for construction projects ranging in size from a million dollars to hundreds of millions of dollars. Morse/Diesel has managed the construction of projects such as Chicago's Sears Tower, the Grace Building and the Pan Am Building in New York; renovations at Avery Fisher Hall and the New York Stock Exchange; institutional building for The National Gallery of Art in Washington and New York's Lincoln Center. Morse/Diesel's primary area of expertise is the management and supervision of commercial and institutional construction projects such as hotels, apartments, jails, hospitals, corporate headquarters, industrial facilities, office buildings, and tenant work. Morse/Diesel has not been involved with large power projects, road work, or large industrial complexes. However, in 1981, the total value of projects under contract exceeded 1 billion dollars.

Contract procurement for the company is accomplished through negotiations with potential clients. As a constructon manager Morse/Diesel's policy is to avoid the potential conflict of interest which may be present when managing both negotiated and hard bid projects in the same limited resources marketplace. Morse/Diesel does not bid work. Morse/Diesel's services are reimbursed on a fee basis, similar to the arrangements for architects and engineers. Therefore, client references on the company's past performance are crucial to its continued ability to secure new work.

Morse/Diesel has offices in nine states: New York, Massachusetts, Florida, Maryland, Illinois, Kentucky, Wisconsin, Minnesota, and California. The company is organized around the individual offices as separate profit centers, however, the company is divisionalized along project type guidelines as Commercial, Hospital, and Industrial divisions wherein staff specialists and senior management concentrate. The company management is controlled by 33 officers. Management committees composed of these officers make and implement company policy. Standing committees are currently: Executive Committee, Operations Committee, Standards Committee and the newly formed Computer Committee.

EXPERIENCE WITH SMALL COMPUTERS IN CONSTRUCTION AT

NORTH DAKOTA STATE UNIVERSITY

By Merlin D. Kirschenman,[1] M. ASCE

Galen Nation[2]

ABSTRACT: The Construction Management and Engineering Department at North Dakota State University is assisting the regional contractors in developing the necessary tools to improve management of construction projects using two IBM-PC computers. Because of the capacity and low costs of these small computers, it is now economically feasible to have one at each project office. The intent is to mesh the field level personnel's ability to use computers with ease of use of the small computer. The result will be more effective use of this tool. Software to provide more appropriate management information systems need to be improved. Also software to tie specification writing, scheduling, estimating, cost engineering, and interim reports on a standard code of accounts need to be developed. These computers are used to teach our students in programming and computer solutions of problems. Also these computers are used to show contractors how they could improve their operations. The appropriate use of the computer as a tool will significantly improve the performance and production of all construction industry participants.

INTRODUCTION

I.   Need for Microcomputers

   A.   Cost Effectiveness of Data Processing and Computations

     No, microcomputers are not reproducing themselves and crawling into offices and classrooms! They are being purchased and carried into most places of business, more specifically in the construction industry. These executives are buying microcomputers and software the best way they know how to implement into their company for various purposes to eventually save time and money. Microcomputers within construction and engineering are becoming quite cost effective for data processing and computations after the intial investment of time and equipment. This can be verified by considering an example.

     A data processing system was set up by a local consulting engineer for specifications writing from master specifications.

---

1. Assoc. Prof., Department Chairman, Construction Management and Engineering Department, North Dakota State University, Fargo, ND
2. Assistant Professor, Construction Management and Engineering Department, North Dakota State University.

It was observed by the consultant that specifications could be produced in 40 percent of the time required for hand edited specifications.[2] A weeks salary for a chief specifier and data processing secretary is over $800. Therefore, over $480 per week in salary is saved in writing specifications. The savings is compared to the cost of a system, both hardware and software.

A typical computer system consists of minimal hardware, such as microcomputer with double disk drive and letter quality printer for under $4,000. Typical software consists of a master specification and a text editing program for approximately $1,600.

The approximate combined cost of $5,600 could be recovered in 12 weeks by saving $480 per week. In addition, 4 weeks of start up time is necessary. Therefore, in approximately 16 weeks the entire system would be paid for. Return of this type is contingent upon steady usage. The savings on engineering calculations by computer versus hand computations would be comparable.

B.   Adoption by the construction industry in North Dakota

Most of the construction companies in North Dakota are relatively small and they do not use very sophisticated management control methods. Consequently the adoption of computer methods has been very slow. The use of computers in construction has been primarily for accounting type functions.

C.   Familiarity of Small Computers Necessary for Students

Because of the capacity, the economics, and the high probability that these microcomputers will work their way into every construction project office; students graduating from construction programs should have the capability to program and use these computers. The current thinking of many people familiar with the capabilities of these computers is that they will be the important working tool for the next several years.

II.   Evolution from Main Frame to Microcomputers

A.   Extensive Usage of the Main Frame Computer

As essential computer hardware for the Construction Management and Engineering Department at North Dakota State University, the university main frame computer has been used extensively . It has been used for complex computations all the way down to correcting tests. To some extent it has been overused, especially for student assignments and computer games. The main frame computer has been and will continue to be an effective tool for complex computations, such as in the area of research by most professors. In recent years, microcomputers have become available at affordable prices to handle the smaller computations previously forced upon the big university computer.

B.   Current Usage of Microcomputers

In North Dakota, microcomputers are being used or being considered for construction and engineering. Most companies are small in size which makes justifying a change much harder. Training of contractors and engineers can be of concern for those old enough to not have received computer training in school. Many of these individuals have not yet made the transition and will require specialized training. For the companies that are using microcomputers which are fast becoming the majority, they are finding out that these computers are cost effective. A small percentage have not been able to make the transition primarily due to a lack of training.

A few of the larger companies have been operating with a main frame computer for several years. The microcomputer is an asset to those existing systems. When it is tied into the mainframe, it serves as a computer terminal in addition to a microcomputer. Field operations are run by the microcomputer, which can be tied into the big computer at the main office. The project manager in the field feels like the project is completely managed by him. This manager immediately finds cost or scheduling information with his microcomputer and provides updates.[1] Therefore, microcomputers are providing benefits in small as well as large construction and engineering firms.

III. Feasibility Study for Types of Small Computers

A.   Study of Hardware Available

The Construction Management and Engineering Department and the College of Engineering and Architecture made an indepth analysis of microcomputer availability. The basic criteria used to judge and compare the different brands was: (1) computing capacity, (2) costs, (3) expandability, (4) the availability of software for the unit, and (5) a judgement of the probability that the computer manufacture would be in business a few years down the road. To assist us in this effort, a consultant was hired who was familiar with the capabilities of several brands, had designed and built several similar microcomputers for his corporations internal use, was not selling computers, and offered a very high probability of presenting an objective viewpoint. The conclusion drawn from all this research and analysis was that the IBM-PC was becoming the standard and the availability of software would be predominantly for the IBM-PC model. Therefore the recommendation was to buy a microcomputer that was compatible with the IBM-PC. At the conclusion of this research in April 1983, the IBM-PC compatible microcomputer that appeared the most appropriate was the Columbia Personal Computer. In the interests of standardization, North Dakota State University preferred to go with the IBM-PC, even though it had less expandability than the Columbia.

IV. Selection of Software

A.   Compatibility of Software

The market for availability of software is in a very fluid stage at the present time. As stated previously in this paper,

it appears that IBM introduced some standardization into the
market with the introduction of the IBM-PC model. Consequently
the writers of software programs appear to be concentrating
on addressing this market and preparing software that will
operate on the IBM-PC. Purchasing of software before the microcomputer is sometimes recommended.

B.   Complexity of Analysis and Ease of Usage

There are two general catagories in which microcomputer
software are classified. The first one consists of software
that incorporates fewer computations and simplification of
operations. This type promotes ease of usage and allows
starting computer operators to become familiar with the system.
The capability is limited to fewer problems of somewhat reduced
complexity.

The second catagory of software consists of very complete
packages, which require considerable quantity of computer input.
This type promotes usage by thoroughly trained computer operators.
The capability is extensive and provides computations for complex
problems. Beginning computer operators generally start with
catagory one software, then advance to catagory two software
packages.

C.   Search for Software for Management Information Systems

Our search of the commercially available software that
could tie together scheduling, estimating, cost engineering
and management information systems has led us to "Constructors
Inventory of Computer Software for the Construction Industry"[3]
and Construction Computer Application Directory.[5] North
Dakota State University is in the process of obtaining
information on these programs that appear adequate. If
necessary the Construction Management and Engineering Department
will write the needed software as resources become available.

V.   Updating Faculty and Staff for Use of Microcomputers

A.   Hands-On Seminar

Introduction to microcomputers can best be accomplished
by basic, simplified instruction. Hands-on instructional
seminars consisting of small groups are valuable in starting
the learning process. Each person is placed at a microcomputer
and instructed in logging on and simple operations. Use of the
hardware and trouble signs are part of the instruction. Use
of back disks in simple procedures is included in the seminar.
The seminar taught at North Dakota State University did not
enter into specific software applications.[4]

B.   Additional Training

Other seminars might be offered by specific groups or

companies that will all be using a particular software package.
Again, this seminar should be hands-on with particular problems
relating to the people involved. These simple problems solved
by each individual should relate to the more complex problems
that each one will learn to solve on their own.

C.   Practical Applications

To complete the learning process, the training gained in
seminars must be used in practical problems. Individual study
is necessary to apply the seminar information to a typical
problem of construction. The more complex problems in practical
application will bring about confidence and solidify the use of
microcomputers for the individual.

VI.   Training with Departmental Microcomputers and Software

A.   Student Course Work

Students in the Construction Management and Engineering
Department are acquiring computer skills both in Computer
Sciences and this department. The majority of courses offered
by this department will soon be using microcomputers. The ABET
accrediting agency and the construction industry have stressed
the importance of a transition to microcomputers. The change
has taken place slower than expected primarily due to the lack
of funds.

The software acquired for student usage is of the first
catagory for ease of learning. The sophistication of these
packages is of a low level but adequate for instructional
purposes. Students may pursue the topic in senior or graduate
projects. In these, they may advance to the second catagory
for more complex software.

B.   Senior or Graduate Student Projects

Students after or near graduation are encouraged to go into
more individualized study and projects. Undergraduate courses
are expanded upon. In addition, application to the construction
industry is necessary. We encourage cooperation with local
contractors and engineers. Promotion of excellence in education
will help academia to stay current in technological advancements
and if possible to get ahead.

VII. Conclusions

Microcomputers are cost effective in construction and
engineering if properly implemented. In North Dakota small
contractors are slower than other areas to employ the advantages
of microcomputers. This is due to both initial hardware and software costs but primarily due to training and upgrading of
personnel. In academia, faculty and students must be properly
trained in computer aided design and management and be able to
bring computer aided construction and engineering to the construction industry, specifically to small firms in North Dakota.

BIBLIOGRAPHY

1. Chidester, B., "Microcomputers-Providing Low-Cost
    Solutions to Management Problems," Constructor, Dec.
    1983, 1957 E. St. N.W., Washington, D.C. 20006.

2. Cowman, W., Foss and Associates, 1701 S.W. 38th St.,
    Fargo, North Dakota 58102.

3. Snow, D.B. Ed., "Constructors Inventory of Computer
    Software for the Construction Industry," Constructor,
    Dec. 1983, 1957 E. St. N.W., Washington, D.C. 20006.

4. Sprafka, A., "What is DOS Anyway?" Computer Center
    North Dakota State University, Fargo, North Dakota
    58105.

5. Construction Computer Applications Directory, 1105
    F Spring St., Silver Springs, Maryland 20910.

Construction Management Software Requirements

By Thomas R. Anderson,[*] M. ASCE

Abstract: With the rapid improvement in capabilities of microcomputer systems, it is possible to develop software for small computers previously only possible on mini or mainframe computers. This paper will examine the software requirements for construction companies and provide a guide for integrated software design and selection on small computers.

Introduction

A computer system for business management must be designed to serve the requirements of both accounting and management information. For accounting, accuracy and ability to audit or prove each transaction (audit trail) are the most important. For management, timeliness and relevance of the information are more important than accuracy. Management information systems should also follow the principle of "mangement by exception", but few packaged systems presently do this. Custom reports are therefore needed to meet the exception reporting requirements of each company. The trend in business computer system development is toward packaged software. The development cycle is:

1. Survey the management information needs of the company.

2. Select packaged software that best meets those needs and can be easily modified for custom reporting requirements.

3. Acquire a computer system compatible with the packaged software.

4. Augment the packaged software with custom data storage and reports to meet special management information requirements.

Management Information Software Requirements

When first looking for construction mangement information software, you can be easily bewildered. Remember that selection of a computer system is like any other business decision: you are the customer and have the right to have every question answered about your business requirements before you buy. Take your time to make a decision: many companies have been "burned" by a premature computer purchase decision. You may want to retain a professional consultant to help you define your needs: if he does not represent a specific brand of hardware or software, he will be more objective. A "systems house"

[*]Management and Computer Consultant, 816 N. Stadium Way
Tacoma, WA 98403

is a computer company providing complete hardware–software systems, modifying the software to meet the needs of each customer. If you can find a systems house in your area with good references from other construction business users, you may want to deal initially with them and skip the consulting phase. Dealing initially with a good systems house may save you time and money, if you are sure of your needs. Whether you deal with a consultant or a systems house, learn as much as you can about computer systems, software and the needs of your business. To assist you in your decision, the following section describes software requirements for construction management information systems.

Integrated Construction Management Software

Good accounting and management information software is integrated: entry of source data is made only once with all related transactions updated automatically. With integrated software, payroll transactions to the general ledger and job cost are automatically kept in balance. Figure 1 shows the relationship of applications in an integrated construction management information system.

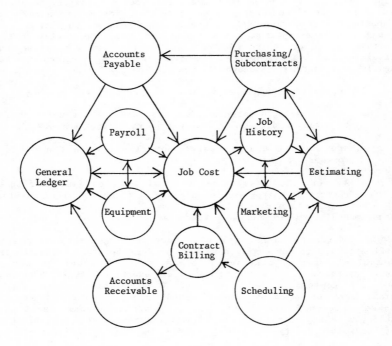

Figure 1. Integrated Construction Management System

Traditional construction accounting systems included only general ledger, accounts payable, accounts receivable, payroll and job cost. These applications were not always fully integrated: payroll did not always balance to job cost, data on material purchases and charges to jobs were often inaccurate or late. Many journal entries and corrections were necessary before monthly financial statements were produced. Due to the many corrections, the income statement could not be produced directly from the computer system. Integration of the above applications minimizes the number of adjustments which must be made manually and insures a regular flow of accurate information from the computer system. To complete the management information system shown in Figure 1, the applications of equipment, purchasing/subcontracts, contract billing, scheduling, estimating, marketing and job history are added. These additional applications make it possible to encompass the management information needs of all construction business functions in the computer system.

To follow a job from start to finish in Figure 1, imagine a potential job starting as a lead in marketing. As more information is learned about the lead, the job history file is consulted by the Marketing Department for information about past similar jobs performed by the Company. The efforts of the Marketing Department are successful, and a request to bid the project is received. Estimating prepares a bid for the project, using the computerized estimating system, job history and standard estimating codes and costs developed by the Company. Scheduling prepares a preliminary schedule for the bid. The job is awarded to the Company, and Estimating, Purchasing and Subcontracts prepare a complete takeoff with bills of materials and scope of work for subcontractors. All bills of material and draft subcontracts are prepared on-line to the computer system (on-line means that the text and data is stored on a disk, instantly available for editing on a video screen). Discrepancies between the bid and contract bills of materials are reviewed by Estimating and Project Management: Was there a mistake in the estimate? Is there a change in scope for which we could be reimbursed? The bills of materials and subcontracts are then edited on-line and purchase orders and subcontracts are automatically produced from this information. The purchasing system keeps track of purchase order commitments for each line item (cost code). Total material costs and the status of material purchasing are thus always known. Processing of accounts payable is easier with the purchase order detail already on-line. Project Management now evaluates progress for the first contract billing. With every line item of job cost identified as to contract pay item, it is possible to automate the contract billing. Through field audits of progress, information about quantities put in place by line item, and schedule information, it is possible to document progress to the customer. Using this information, the Contract Billing System prints out an invoice and certificate for payment in a format suitable for approval by the customer. As the job progresses further, Project Management is able to follow the status of each line item of the job by percent complete, quantity installed, actual cost to date and projected cost at completion. Records are kept for each piece of equipment for job costing and for maintenance and depreciation. Change orders are kept separate from the base contract for accuracy in progress reporting against the estimate. At completion of the job, information is available for every line item to support any claims for extra work. The job is then filed in the Job History File for reference in bidding future jobs.

The system shown in Figure 1 is still an ideal. Few running systems have the scope and integration of this system. Different types of contracting require a change of emphasis in the software applications required. Mechanical and electrical contractors will have greater emphasis on purchasing and labor cost control. General contractors will have greater emphasis on subcontract administration. Heavy construction and excavation contractors will emphasize equipment cost control. Each construction company has unique management reporting requirements. How can you satisfy the need for custom reports with a packaged software system? There are two problems:

1. The data required for the report is already entered and stored by the system, but there is no way to generate the custom reports without extensive programming.

2. Additional data must be entered and stored by the system to meet the reporting requirements and there is no way to modify the data files.

It is difficult to solve these problems. Much microcomputer software is sold in a protected form: the original programming language (usually BASIC) has been converted to a form readable only by the computer. Modification of protected software can only be accomplished by the original programmer or systems house. It is unwise to acquire software in protected form unless it is supported by a local system house that is likely to stay in business and has the original programming (source code). You may be better off if you have the source code, but even source code is useless if it is not well documented both by user manuals and comments within the program. Well written programs will require less documentation. Be sure of your contractual rights to computer software. Packaged software is generally sold as a one-time license. You thus have the right to use the software, generally only one one computer and do not have the right to resell the software. If you require much customizing, get the systems house to provide source code and documentation for the custom programming. You may also want to restrict the systems house from using any unique features of your custom programs for their other customers.

If you decide to have an employee write software, remember that it is difficult to control the cost of computer software development. In "The Mythical Man Month", Brooks (1) cites programmer productivity variations of 10:1 ! He also discusses the optimism of programmers which leads them to drastically underestimate costs and schedules to produce custom software. Due to the uncertainties of in-house software development, you may want to consider retaining a computer systems house to develop and maintain your hardware-software computer system. You should approach the decision to retain a systems house in the same manner you would select an attorney or certified public accountant and expect the same professionalism from them. Finally, you should have a computer plan and periodically review the plan.

Computer systems require periodic changes and enhancements which should be done according to the plan.

Data Base Concepts

Like much computer jargon, the words "data base" are used by many but understood by few. In "Everyman's Database Primer", Byers (2) writes: "A database is a collection of information organized and presented to serve a specific purpose." The data base concept is presented here as a means of improving the flexibility of processing in a construction management system. Data base methods can be used to describe all of the data in a computer system such that the data is completely independent of the applications programs that process the data.

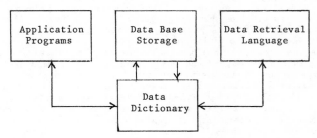

Figure 2. Data Base Organization

In Figure 2, the elements of a data base system are described. By making the data base storage independent, data structures and programs can be easily modified. Special inquires and reports can be made directly from the data retrieval language using simple English-like commands. The data dictionary stores the names, definitions and characteristics of all of the data items used in the data base. The data dictionary maintains consistent use of data by the many application programs and by the data retrieval language.

Data base driven computer systems will be of great benefit to construction management. Each new project creates a demand for special reports which cannot be satisfied by existing programming methods. Using the data retrieval language, project personnel can run their own reports with a minimum of special training. Information bottlenecks and need for central office data processing staff will be greatly reduced. Strong central office planning and support are still required to maintain integrity and security of the data base.

There are a number of database languages available now for micro-computers. I am not aware of any complete microcomputer applications for construction management written in a database language that meet the requirements stated in this paper. IBM'S Construction Management Accounting System (CMAS) (3) for the System/36 minicomputer is an example of a system with a data retrieval language interface.

Implications of Multi-User Systems

The first computers processed only one job at a time. Large computers can now serve hundreds of users at one time. Due to their limited memory and speed most microcomputers process only one job at a time. Multi-user microcomputers can serve four to eight users through use of either multiple processor or multiple task architecture. Multiple processor is the most prevalent architecture and dedicates a complete "computer on a board" to each user, sharing only a central printer and hard disk facility. Multiple tasking systems divide up the memory into partitions, each user having one partition. Users are given time slices during which the central processor executes a small portion of their task. Multiple tasking is not viable for present microcomputer systems due to the relatively slow processors and small memory sizes. Another limitation on multi-user micomputers is the operating system. The operating system is the software that controls the computer in loading programs, accessing the display terminal, printing and access to disk storage files. Microcomputer operating systems are presently not adequate for multiple user applications. With the increase in speed and memory size of microcomputers, minicomputer operating systems will be adapted to microcomputers and the multi-user capabilities will increase dramatically. You should investigate vendor claims of multi-user capabilities carefully, including tours of actual installations. Another limitation for multi-user microcomputer systems is the applications software. Most microcomputer software is written for single user systems and processes information in batches. Multi-user systems require security passwords, data base structure, record locking (prevents two users from accessing the same data record) and sophisticated operating systems. These difficulties will be quickly overcome and multi-user microcomputer systems will become commonplace in the construction industry.

Conclusions

Applications software development has lagged behind the explosive growth of microcomputer hardware. Software development is an intellectual process limited by the programming languages and software tools now available. Availability of data base development software and other software development tools for microcomputers will speed applications software development. More microcomputer software companies will pursue the construction market and develop applications software that responds to the needs of construction companies. Multiple user microcomputer systems complete with software will be available in the $10,000 to $20,000 price range. Systems and applications software for microcomputers will be upwardly compatible with minicomputer systems, protecting software investments as a company grows. Construction computer users will form user groups to influence software companies to develop better software for construction.

References

1. Brooks, Frederick P., Jr., The Mythical Man-Month, Addison-Wesley Publishing Company, Reading Mass., 1982, p. 14, p. 31.

2. Byers, Robert A., Everymans´s Database Primer, Ashton-Tate (publishers of dBASE II database language), Culver City, CA, 90230, 1982, p. 2.

3. "Introducing Construction Applications – General Information Manual – Construction Management Accounting System (CMAS) – System/36", IBM Corporation, Atlanta, Georgia 30358-0331, 1983.

Construction Estimating Decision Support Systems

Darryl L. Craig[a] and H. Randolph Thomas[b], M. ASCE

Abstract

A discussion of important considerations in the development of a computerized Decision Support System (DSS) is presented. These considerations include general hardware and software issues relevant to the design of an effective construction estimating DSS environment, including the selection of an appropriate host computer language. Twelve attributes of a quality computer software implementation are proposed and exemplified by the features of an established construction earthmoving estimating package.

Introduction

The construction industry is the the midst of a computer revolution. Computer accessibility is affordable by even the smallest contractor, and new developments that enhance the power and utility of desk-top microcomputers appear daily. The availability of computer software, particularly for payroll, accounting and scheduling, offers benefits that cannot be ignored. For example, the "spread sheet" programs recently developed provide unprecedented number crunching capabilities.

However, estimating programs are less accessible due to the diversity of estimating practices. Generalized estimating programs are probably not attainable in the near future. For some simple applications, the "spread sheet" programs may be the solution. But for more complex problems, estimating programs need to be developed that   are designed for specific construction industry applications.

This paper contains a discussion of various design aspects for the development of a computerized construction estimating DSS. The term DSS, originating from the management information systems area, refers to incorporating models into information system software, providing useful information to support unstructured decision making activities, and furnishing the systems users with powerful yet simple to use software for problem solving. Appropriately applied, DSS principles can substantially improve the construction estimating environment.

[a]Assistant Professor, Department of Accounting and Management Information Systems, The Pennsylvania State University, University Park, PA 16802.

[b]Associate Professor of Civil Engineering, The Pennsylvania State University, University Park, PA 16802.

The discussion begins with a review of relevant characteristics of the construction estimating environment, followed by general hardware and software considerations. A review of the APL computer language as the language of choice for DSS implementation is summarized with twelve specific attributes of a quality software implementation. Finally, a synopsis of an established construction estimating program (SEMCAP) written in the APL language is presented.

Nature of the Construction Estimating Environment

The first decision for a contractor is to determine if a computerized decision support system approach is justified. Size of the firm, volume and type of business, and availability of adequate software are but a few of the important factors to be considered. While manual calculations may be adequate, computers can offer speed and accuracy not otherwise available. In addition, the system can generate alternatives and perform sensitivity analyses. With the lowering of timesharing rates on mainframes and a highly competitive microcomputer industry, computerization is be more readily justified.

In the past, computerization implied the need for a computer operator and/or programmer. Turnaround time was anywhere from several hours to several days. For certain aspects, such as payrolls, this may still be satisfactory; however, for the estimator, turnaround must be immediate. The estimator must be able to work in his own environment which involves the use of manuals, plans, specifications, and other documents. The use of the computer must be oriented to rapid decision making, not just on producing a product at the end of the week. Indeed, this aspect is one of the central issues in the development of decision support systems. The dynamics of the estimating environment are such that estimates must be produced using a CRT terminal (or microcomputer) at the estimator's desk without the need for a programmer or operator. With a mainframe, the system should be a timesharing one. Because estimating data bases are relatively small, microcomputers seem equally applicable.

General Hardware Considerations

The suitability of the hardware environment is of primary consideration. Computers with slow response time (problems associated with too many simultaneous users or slow processing speeds), unfamiliar keyboard layouts (usually associated with microcomputers), and limited general software (operating systems, computer languages, and utility programs, and data base systems) absolutely restrict the capacities of any system development. When a suitable hardware environment is being considered, bigger is not necessarily better. Many microcomputer configurations offer ideal environments for decision support system development and installation.

A microcomputer system for decision support system applications should include a typewriter style keyboard, an 80-column screen display, plentiful software availability, peripheral devices (printers, plotters, etc.), adequate memory, two disk drives, and local support services. While decision support systems can be developed on more limited hardware and software, the results may be unduly primitive and ineffective.

General Software Design Considerations

Several general software design principles require consideration. Simpson (10) presented the following important ones: 1) provide feedback, 2) be consistent, 3) minimize human memory demands, 4) keep the program simple, and 5) match the program to user's skill level. While these considerations seem obvious, they are not simple to apply. A large number of commercial programs do not perform according to these principles.

The first consideration, providing feedback, can be implemented in several ways. For instance, data entry requires some system acknowledgment of acceptable data entry and checking for erroneous data entry. Feedback should also include error messages designed to provide the user with definitive and complete information. An error number with cumbersome documentation support is not sufficient. An on-line user assistance or HELP capability is very important. In addition, the system should provide information regarding its progress, especially when long calculations are being conducted. This type of feedback helps assure appropriate user response and aids in user confidence.

Simpson's second consideration, consistency, also supports user input by providing a standard format. Data entry, menus presenting alternative responses, and user instructions presented in consistent formats are more easily recognized and reduce ambiguity. As a user becomes familiar with the format style, efficiency with the system increases. This efficiency is particularly evident when new portions of the system are applied.

The third consideration, minimal memory demands, implies that an effective system maximizes the use of computer memory to accomplish the calculations and does not rely on the less reliable human memory. The remaining considerations also accommodate the user. By keeping the system simple and at the user's skill level, the mission of the system can be performed most effectively. Complex systems that offer multiple alternatives and require many user decisions, detract from the overall performance of any decision support system. Systems that support different levels of user skill allow a novice user to work with a minimal and simple system and a more expert user to work with a more powerful and complex system.

The APL Language

APL (A Programming Language) was developed as a mathematical syntax by Kenneth Iverson in the late 1950's, but was not implemented as a computer language until 1966. It is an extremely concise, general purpose computer language that can be easily learned by non-computer scientists. The language has been implemented on most large mainframe computers and many microcomputers.

While a complete analysis of APL is not feasible, the major advantages and disadvantages of the language for decision support implementations are presented below. The advantages are:

1. Simplicity of Syntax - the lines of code in APL are logically consistent. While the language has many powerful features (each represented by a symbol), the additional power that results more than compensates for the terse notation.

2. Mathematical Notation - APL is very popular for solving problems that perform sophisticated mathematics. The notation is clear and the problem can be reduced to a controllable size.

3. Matrix Handling - APL can perform operations on multiple dimension matrices. For instance, to add two matrices you simply insert a plus (+) symbol between the two variable names.

4. Interactive Nature - APL commands can be executed by either typing the command or putting the command on a line in a program (called functions).

5. Powerful Primitives - Primitives are operators like plus and minus. In APL, there is a long list of primitives. For instance, one primitive can solve a series of simultaneous equations or invert a matrix with a single character command.

6. Partitioned Software - The programs (functions) in APL are separate entities that coexist in the workspace of computer memory. As a result, each function is maintainable as a separate program. An APL implementation consists of many functions interacting in an environment called a workspace.

7. ERROR Processing - In APL, data entry and processing errors can be trapped and processed, a feature that allows production of software that is difficult to "lock-up."

8. Maintainability - APL facilitates software maintenance and enhancement. Only the functions that require change need to be reviewed. In other languages the entire program must be reviewed, a costly process.

9. Software Development Productivity - In APL, due to its inherent efficient nature, it is easier to program a task than in other languages. It is estimated that each line of APL is equivalent to 15 lines of COBOL. APL programmers have been observed to produce 7-17 lines of debugged code per day whereas COBOL programmers produce 10-65 lines of code per day. Given that 7 to 17 lines of APL are equivalent to 105-225 lines of COBOL, APL programmers are approximately 2-3 times more productive than their COBOL counterparts.

Attributes of a Quality Software Implementation

     Unless software producers recognize the type of user in designing
a program, it is likely that the program will be only marginally
successful. The estimator may be a college graduate or may only have
a high school diploma. If over 30 years of age, he will probably have
had little or no computer exposure. In either case, there is only
minimal time to learn the program. Above all, it must be easy to use.

     In the absence of large data bases or complex calculations,
interactive programs are almost always more desirable than batch
programs. Yet, there is a broad spectrum of interactive programs.
Batch programs can be converted to interactive ones by adding prompt
statements for data entry. However, these generally lack the
sophistication needed to be truly effective. Listed in the following
paragraphs are some of the desirable characteristics of interactive
programs. Programs incorporating many of these features can be
considered as being sophisticated interactive programs.

     1. Simple Job Control Language (JCL) - The JCL should be simple
and easy to apply. The commands should be descriptive of the
instructions being provided, e.g., off, load, print, etc. The fewer the
commands, the better. The use of codes should be avoided in favor of
alphabetic commands. There are far more frustrations over how to
access and use the computer or program than there are over how the
program works.

     2. No Black Box - Estimators must be able to develop confidence
in computer programs. He should be able to verify all the calculations that are being made. As much as possible, the calculations and
sequences should be identical to that which the estimator would do
manually.

     3. Leave Decision Making to the Estimator - Programs should be
designed to provide guidelines instead of making decisions. Decision
variables and parameters from tables and equations should not be used
without the approval of the estimator.

     4. Built-in User's Manual - As much as possible, the user's
manual should be incorporated into the program. The user should be
able to access descriptive information via help commands to assist with
data entries, system commands, and error messages. The experienced user
should be able to avoid these altogether.

     5. Flexibility - The user should be able to get in and out of
various program functions without destroying previously entered data.
The roadmap through the estimating process should allow the user the
maximum freedom to proceed as he chooses.

     6. Ready Access to Other Data - Preparatory calculations such as
earthwork quantities and haul-mass diagrams should be available at all
times.

7. Sensitivity Analyses - Evaluation of alternative schemes should be possible without having to reenter all the data. Time is of the essence.

8. Easy Correction of Data Entries - Input errors should be easy to correct and subsequent calculations automatically updated.

9. Easily Understood Error Messages - Help commands should aid in understanding these. The ability to isolate an error should be enhanced.

10. Formal Checks for Data Entries - Checks should be made to preclude the incorrect entry of the wrong number or type of data.

11. Hardcopy Printout - Detailed summary printouts should be possible.

12. Graphical Capabilities - These should be available at the terminal or in hardcopy form.

The Disadvantages of APL are:

1. Keyboard - Since APL uses many special symbols, the standard ASCII typewriter keyboard is unacceptable. Likewise, special terminals are required to display and print the character set. With the advent of matrix printers, graphic terminals, and ink jet technology, this problem has been reduced.

2. Cost - Unlike the basic language, APL must be purchased separately for a microcomputer. Prices range from $200. - $600. However, given the power of the language, the cost is easily justifiable.

3. Memory - Because APL has many primitive functions, it requires more computer memory than other languages. The cost of computer memory has declined to where this is no longer a major problem.

4. Processing speed - Like basic, APL is an interpretive language. As a result, it does not process as fast as compiled languages like Fortran or COBOL. For most decision support applications, processing speed is not a limitation.

5. Unlearning - Programmers trained in other languages are often uncomfortable in an APL environment. Old habits are hard to break.

In summary, APL is a feasible alternative for DSS development. Given the scarcity of computer programmers and the decreasing cost of computer hardware, APL is increasing in use. The advantages far outweigh the disadvantages.

SEMCAP - A Case Study

An estimating program entitled SEMCAP (Systematic Earthmoving Cost Analysis Program) was written in the APL language. It is used herein to illustrate some of the sophisticated capabilities that can be incorporated into a decision support system for estimating.

SEMCAP is a deterministic estimating package for calculating the cost and duration of various earthmoving operations. The basis for the calculations pertaining to scraper, ripping, loading, and hauling operations are manufacturer's handbooks. The details presented by Caterpillar, Eclid, International, and Terex have been combined into a single procedure in such a way that the estimator may use a particular manufacturer's recommended procedure or may develop a hybrid procedure using various steps from several sources. The drilling and blasting operations are based upon a comprehensive deterministic procedure developed at The Pennsylvania State University (13). There is ample opportunity for the user to incorporate his own judgment. The computer performs the calculations; the estimator decides on a course of action. Because the procedures in SEMCAP are the same as what the estimator does manually and the computer provides guidance, not decisions, there is no "black box" mystic created.

Figure 1 is a representation of the SEMCAP workspace. First notice the simplicity of the system commands (JCL). They are few in number and are readily understood. No coding is necessary. In APL, there is no distinction between main programs and subprograms. They are all referred to as functions. To initiate a function, the estimator need only type the function name. For example, to estimate ripping cost, the estimator types RIPPING. He is then prompted for the initial data entry.

A major portion of the user's manual is incorporated into SEMCAP. The prompt statements are clear and concise, including units. Data checks are included throughout to ensure that the data entries are of the proper type and number. If the user encounters difficulties, he may call various information functions for help. These include descriptions of the workspace, user options, system commands, and error messages. The user may ask to see tables that contain technical information, such as scraper load times, acceleration factors, front-end loader cycle times, and numerous others. Because these are functions, only the name of the table needs to be entered.

Program flexibility is one of the key features of SEMCAP. The various functions can be exercised in any order desired without the necessity of using menus. The user may also exit programs in the midst of performing estimating calculations. For instance, suppose in the process of calculating drilling and blasting cost, the estimator needs to retrieve some other information stored elsewhere. By typing OUT, the estimator may perform a loader estimate, determine the value of a particular variable, correct a data entry, review a table, or perform any other task or operation available in SEMCAP. Typing DRILLBLAST places the user back into the drilling and blasting function at the same location where he exited. All data previously entered is

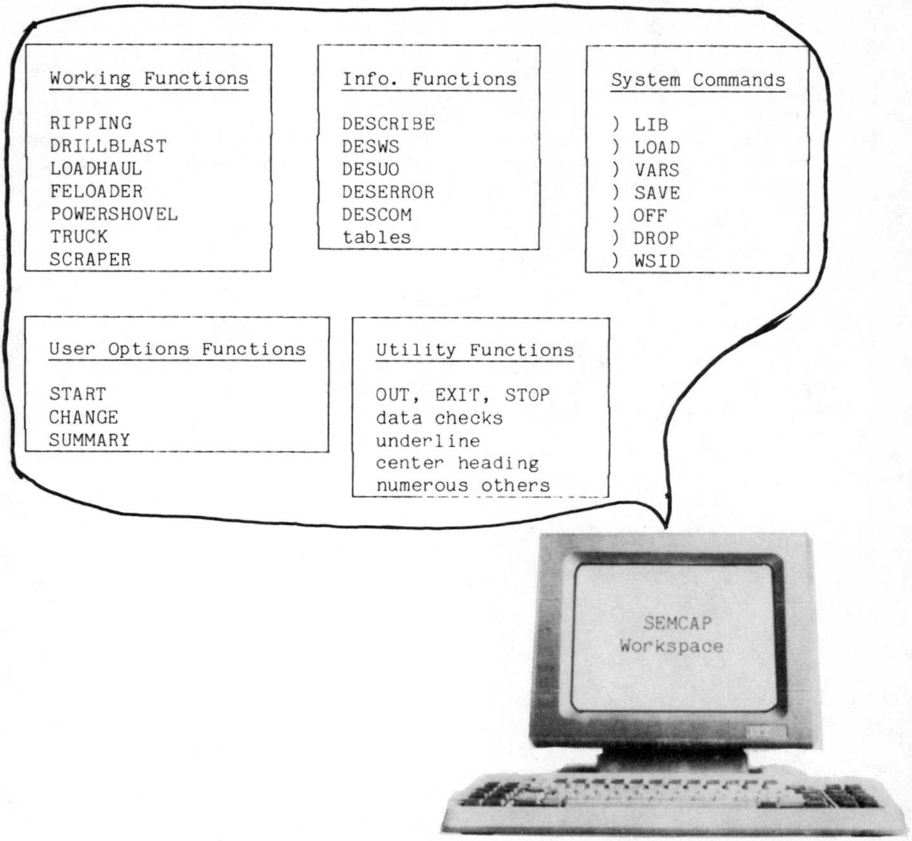

Working Functions	Info. Functions	System Commands
RIPPING	DESCRIBE	) LIB
DRILLBLAST	DESWS	) LOAD
LOADHAUL	DESUO	) VARS
FELOADER	DESERROR	) SAVE
POWERSHOVEL	DESCOM	) OFF
TRUCK	tables	) DROP
SCRAPER		) WSID

User Options Functions	Utility Functions
START	OUT, EXIT, STOP
CHANGE	data checks
SUMMARY	underline
	center heading
	numerous others

SEMCAP
Workspace

Figure 1 - SEMCAP Workspace

retained. Further, if data entries have been corrected or changed, all
subsequent calculations will be automatically updated. This ready
access to other data and the ease with which data entries can be
changed means that sensitivity analyses and alternative schemes can be
rapidly examined.

Output can be obtained in summary form from the CRT or in hardcopy
form. Executive summaries are available at the CRT, whereas detailed
summaries are routed to a printer.

Conclusion

Decision support systems for construction cost estimation have
many facets. The choice of what should be computerized, the hardware
environment, software selection, and the program development process
must all be considered. Many elements to producing a quality dds have
been highlighted. It is concluded that the APL computer language is an
effective language to use for DDS development.

References

1. Dawkins, G. "A Language for Everyone," Microcomputing. 1982. pp. 46-52.

2. Haas, Wolfgang. "Standardized Input Conventions for Engineering Software," Proceedings of the International Conference on Computing in Civil Engineering. 1981. pp. 1169-1177.

3. Hodge, Charles S. "Support for Remote Office Computer Users," Proceedings of the International Conference on Computing in Civil Engineering. 1981. pp. 687-695.

4. Jonatowski, Jean-Jacques, "User Conveniences in Application Programs," Journal of the Technical Councils of ASCE. Vol. 105, December, 1977. pp. 43-49.

5. Jordan, R. C. "A Language on the Rise," Computerworld. 1982.

6. Love, Franice E. A Systematic Earthmoving Cost Analysis Program. The Pennsylvania State University, University Park, PA. 1982.

7. Maysforth, G. R. "On APL and Productivity," Communications of the ACM. Vol. 24, 1981. p. 478.

8. Nelson, Brian W. An Interactive Computer Program for Estimating Construction Compaction Operations and Profile Analysis for Earthmoving. The Pennsylvania State University, University Park, PA. 1983.

9. Shoor, R. "Why Doesn't Everyone Use APL? Expert Asks," Computerworld. 1982.

10. Simpson, H. "A Human-Factors Style Guide for Program Design," BYTE. 1982. pp. 108-132.

11. Stoll, Herbert K., and Walter Stensch. "Providing the Computer Resource," Proceedings of the Interaction Conference on Computing in Civil Engineering. 1981. pp. 695-701.

12. Thigpenn, James A. "Microcomputers in Small and Medium Size Firms," Proceedings of the International Conference on Computing in Civil Engineering. 1981. pp. 108-119.

13. Thomas, H. R. and Jack Willenbrock. Planning and Cost Estimating Procedures for Drilling and Blasting Operations. Department of Civil Engineering, The Pennsylvania State University, University Park, PA. 1984.

14. Tomita, Michiya. "The Construction Management by Small Business Computer," Proceedings of the International Conference on Computing in Civil Engineering. 1981. pp. 568-573.

Computerized Estimating, What's Right For You

By Laurence C. True
Demand Construction Services, Inc.
Englewood, CO

INTRODUCTION

Anyone who has ever been an estimator knows that estimating is part science, part art and part arithmetic. The science part comes as a result of our education. We are taught to think things out logically and to consider every decision very carefully. We are also taught to organize and plan our work and to prepare clearer and concise presentations.

The art results from our sense of good judgement and general experience in the construction industry. We know that few people could fill our shoes when it comes to determining the value of work and the proper methods for the construction we are undertaking. This is part of our unique industry that no one can take away from us.

The arithmetic part of estimating, however, takes almost no talent, is very tedious and could be done by almost anyone within our organization. For that matter, it could be done by almost anyone with minimal office skills who owns a calculator and knows how to write. But it is this area of arithmetic that consumes most of our time and causes most of our mistakes in the estimating process.

It is this desire to get rid of the tedium of estimating that leads us to consider in investing in a computerized estimating system. Virtually every contractor is currently seeking or has recently purchased a computer system to help with estimating. It ranks as one of the most heavily sought applications on a computer system today. But there are distinct things that it can and cannot do.

A computer cannot help with many day to day tasks involving "think-time" or "talk-time." A computer system may be able to help us somewhat in the quantity survey, but that aspect depends greatly on the program

72

and varies widely from one program to another.

Much of an estimator's time is spent on tasks involving writing and calculating. These tasks include things such as listing the same descriptions over and over again or tabulating costs of equipment, labor and stock items each time they are used. Tables and estimating manuals must be updated for all the estimators so that each one has current information with which to prepare bids. These are all tasks which can be performed effectively and economically by a computer system.

There are right and wrong ways to select a computer system and there are many systems on the market which may or may not fit your needs, depending upon exactly what your needs are. There are many things you can do to ensure the success of any computerized estimating system.

DETERMINING YOUR NEEDS

In order to assess the potential effectiveness of a computer in your organization, you must first determine the benefit that you expect from computerizing. One could argue that a single bid mistake of significant magnitude could pay for a system. However, it could also be argued that improved procedures, rather than a computer, could eliminate that bid mistake and thus eliminate that offsetting benefit. Thus, in order to make a truly analytical analysis of your computerized estimating investment, a more scientific approach is needed.

First, analyze the people in your organization. Are they generally well organized and receptive to the prospect of computerizing? Computers are of little value if the organization is filled with people who are afraid of computers or who are reluctant to accept computers as a tool. Computers are also of little benefit where there is no overall organizational structure or consistent estimating practices within the organization. If the procedures are bad, a computer will not help. In fact, a computer might make things worse as many people would rather switch than fight. In essence, many old time estimators especially, feel that they have been preparing estimates by their well established but somewhat inconsistent methods for many years and getting by quite nicely. Why should they need to change now?

The answer to this question is obvious from a management standpoint. Some organizational structure and methods improvement is a must if a firm is to stay profitable in this highly competetive market. Thus,

management must accept it's obligation to do a sales job
on the estimating department prior to purchasing a
computer system. The estimators must be ready for it
and accept it.

Secondly, any company considering computerized
estimating must analyze the overall benefit they expect
from a computer system. Clearly, a substantial part of
an estimator's day is spent on activities which cannot
be computerized. Tasks such as calling subcontractors
for quotations, visiting the job site, and discussing
problem areas with colleagues cannot readily be
computerized. Thus, 100% of an estimator's time cannot
be included in the benefit analysis.

To consider computer benefits accurately, only
certain tasks can be included in the analysis. First,
the time spent listing items on a spreadsheet can
usually be improved somewhat as the computerized
estimating system may eliminate a good deal of
repetitive writing. Typical estimating system could save
from 40% to 70% of this time.

Next, the task of looking up prices and writing
them down and then multiplying them times the
appropriate quantities can be almost 100% eliminated by
the computer. This is one of the things the computer
does best. Generally speaking, the computer can assist
somewhat in the quantity surveys. However, this varies
a great deal from one system to another. No hard and
fast rules exist but, depending on the trades involved,
this savings may run from 10% to 50% of an estimator's
quantity survey time. Improved accuracy and elimination
of mistakes may be the greatest benefit in the quantity
survey.

In order to accurately estimate the potential
savings areas and to determine the range of savings
possible from the system, it is advisable to have each
estimator keep track of his or her time for a period
ranging from one week to several weeks. Generally
speaking, the longer the period of time considered, the
more accurate will be the results of the analysis.

At the end of your study period, have each
estimator calculate the total number of hours spent on
each of the identified tasks. Take each of these
numbers and divide by the total hours spent during the
study period to determine a ratio of time for each area.
Multiply these ratios times 2,000 hours to determine the
approximate number of hours spent on each task over the
course of a year. To determine the total potential
savings, mulitply your minimum estimated savings
percentage and your maximum estimated savings percentage
in each time category by the number of hours in that

category.   Add up all the hours for minimum savings and
maximum   savings   and multiply it times the cost of   the
estimator's   wages   to   determine   the   total   potential
savings.   (See   Figure   1   for   an   example   of   this
calculation.)

Estimator: Sue Sharp

Tasks			Time				% Savings	
	Mon	Tue	Wed	Thur	Fri	Total	Min	Max
Site Visit	7					7	-	-
Takeoff	3	5	4			12	10	50
Contract Subs		2	3			5	-	10
Calculate					4	4	80	100
Lookup & List			1	7	4	12	50	80
Think Time	1			1		2	-	-
Misc.		1			1	2	-	-
				TOTALS		40	10	20

Minimum Savings

2,000 Hrs. per year x 25% = 500 Hrs. x $14 = $7000

Maximum Savings

2,000 Hrs. per year x 50% = 1000 Hrs. x $14 = $14,000

Sample of Estmator's Time Analysis
Figure 1

Summarizing   these   numbers   for   your   entire
estimating   department   will provide a range of   benefit
that can be used to determine the relative benefit range
from   acquiring   a computer   system.   Don't   forget   to
include   in your benefit analysis any clerical time that
is spent simply checking arithmetic as that time will be
eliminated completely.

Next,   in   your   benefit   analysis,   consider   the
relative   success ratio of your estimates and   determine
how   many hours are expended on successful   projects   in
retabulating   cost   and material figures   for   budgeting
purposes   and material purchasing.   Computer systems in
their   ability   to   let   you   quickly   and   effectively
reorganize   information   and   to   make   adjustments, can
drastically reduced this time as well.

Once you have determined all the benefits that   may
result from owning a computer system,   you should have a
fairly   good idea of the budget with which you can work.
Basically,   your budget should be evaluated over a   five
year   period of time,   comparing your expenditure for   a

computer system and it's support and maintenance cost to that amount of estimating cost you will save over five years. Be sure to take into account such factors as escalation on wages when considering your five year benefits.

SELECTING THE RIGHT SYSTEM

Next, consider that there are several different types of computerized estimating and that each one has its strong points and weaknesses as well.

First, the most straight forward and simplest computerized method of estimating is that of electronic spreadsheet templates. In this approach, an electronic spreadsheet program such as VisiCalc, SuperCalc, MultiPlan or Lotus 1-2-3 would be used as the fundamental tool. An estimator would prepare computerized worksheets for different types of estimates and different kinds of calculations. These worksheets would be saved on the computer for future use. When a new estimate is being prepared the estimator would reach into a computerized file and retrieve the appropriate worksheet. By quickly and easily adjusting prices and changing quantities or hours as required, an estimate could be recalculated and new revised worksheet could be printed out. Once the initial template is prepared, many different estimates and calculations can be quickly and easily performed.

One of the advantages of electronic spreadsheet estimating is that it is almost identical to a manual estimating system with the exception that the computer, rather than a piece of paper, is used as the notebook. It is usually very easy for an estimator to learn to use this type of estimating method. One of the disadvantages of this approach is that the electronic spreadsheet allows fewer options to the estimator for related tasks such as tabulating of materials for purchase orders. The cost is very low, however, and the training time is minimal in such a system. Furthermore, a very inexpensive computer can be used very effectively.

The total cost for a single user of a microcomputer with a printer and an electronic spreadsheet program would be less than $5,000. Additional time would be required to set up the templates but a reasonably good typist with a little bit of experience can create a new template almost as quickly as a hand written worksheet could be prepared. Thus, the estimator can prepare worksheets as the need arises without actually investing much in the initial setup.

The next popular method of computerized estimating

is that of the price file estimating system. Price file
estimating systems consist of computerized price file
data bases which can be accessed by the takeoff or
quantity survey program.   An estimator lists the items
to be priced by either selecting them from a pre-
determined menu,   entering them with a special keyboard
or by some sophisticated measuring devise such as a
Stylus or Sonic Table.   These types of estimating
programs very often can be purchased with large data
bases built into them and often provide a great deal of
assistance in the quantity survey.   By providing the
contractor with sophisticated means of performing
quantity surveys,   these methods can often provide a
substantial benefit by saving time in the takeoff
process.

        Price file estimating systems,   on the other hand,
are normally best suited to item intensive tasks such as
electrical or mechanical estimating.   In fact, the
larger numbers of programs available for estimating by
price file methods are customized for these trades.
There are, however, available several good price file
estimating systems for structural concrete and drywall,
for example.   Price file estimating systems with large
built in data bases are very often quite expensive in
comparison to electronic spreadsheets.   However, when
one carefully weighs the total time savings of having a
prepared data base, the added cost can usually be
justified.

        The third method of computerized estimating is the
so-called data base/assembly method.   In this method the
estimator can "assemble" certain work packages or work
crews out of a standard item file.   These assemblies can
be used over and over again as required and can be
quickly and easily modified as an estimate is being
prepared.   In this method, an estimator is allowed a
tremendous amount of flexibility in setting up the
system and determining its overall structure.

        While this method seemingly allows the estimator
the most flexibility, it has a substantial drawback
which must be considered.   That is that each item must
be entered into the system by the estimator and
carefully defined so that it can be retrieved again
under some common description.   Likewise, each assembly
must be carefully constructed by the estimator to fit
the widest variety of tasks so that it is ultimately as
useful as possible.

        These estimating systems require the most
forethought in setting up and can be very tedious and
require a great deal of time to make them work
effectively.   While the estimator need not be a
programmer he or she must be able to take a systems

approach to the setup and structure of the estimating system. Without a good organizational structure this system will produce little more than a hodgepodge of prices that, with great effort, can be called upon to prepare and estimate.

The overall worth of a data base/assembly estimating system depends a great deal on the efforts applied by the estimating department. When properly setup, it can be a great time saver and can serve the estimating department well.

When considering data base/assembly estimating, the estimating department must add a considerable cost to the computer system as an allowance for setting it up. This cost must be included in the overall system cost which is to be weighed against the benefits described earlier.

Generally speaking, a data base estimating system is the most expensive but has potentially the greatest benefit for companies that require a great deal of flexibility in the estimating process. Also data base estimating systems are customarily much more useful than other types of estimating systems in that they let the estimator review and analyze his costs many different ways, often allowing complete listing and breakdown of materials and subcontracts for ultimately writing of purchase orders and subcontracts. Again, this can be a great time saver if the contractor is successful in bidding a project.

Data base/assembly can be purchased for slightly under $10,000 for a microcomputer based system. Generally speaking, these systems require a fixed disk and should be run on a computer such as an IBM Personal Computer XT or a Texas Instruments Professional with a fixed disk, for example.

Several of these types of estimating systems are available on large computers and can be used in true multi-user applications where several estimators could be working on different parts of the same estimate simultaneously. In large organizations this is very important and should be reviewed carefully.

MAKING THE DECISION

Considering the various estimating programs that may be available to fit your particular need is not an easy task. The pros and cons of the various systems must be carefully weighed. Also, it is very easy to fall into the trap of looking for the perfect system. There is no such thing as a perfect computertized estimating system. Very often, what is perfect for one contractor is very poor for another. Therefore, when considering a computerized estimating system it is very important to weigh out the benefits and compare them against the cost and pick the system that produces the best cost benefit ratio.

Once the purchase has been made then it is important to make a commitment to its installation and to bringing it on-line quickly. Computerized estimating does not happen without some effort. Price file data bases must be created, and verified. Many comparative estimates must be prepared before the estimating department can feel totally comfortable with the system. Unless a company is willing to make that commitment, then computerized estimating will never by successful.

SUMMARY

In summary, several good computerized estimating programs exist on the marketplace today. Contractors who avail themselves of good systems can generally realize a very respectable return on their investment.

Remember, however, there are no perfect systems and a careful analysis must be made of the software prior to purchasing a system. Be sure that the software governs the hardware selection and not the other way around. There is nothing more frustrating than to buy a beautiful computer and then find out that the software you like won't run on it.

With the increasing power and capability of microcomputers, they are becoming the tool of the trade for computerized estimating. Good software and good methods can make them a very profitable tool. Careful planning along with considering the needs and goals of individual people within the organization will make the computerization of a construction estimating department successful.

Careful selection of a reasonably priced system can bring computerized estimating within the reach of virtually every contractor. The computer is a tool. Take advantage of it.

SUBJECT INDEX

Page numbers refer to first page of paper.

AUTHOR INDEX

Page numbers refer to first page of paper.